Trees in Trouble

STRIEBY

ALSO BY DANIEL MATHEWS

Cascade-Olympic Natural History: A Trailside Reference

Natural History of Pacific Northwest Mountains: A Timber Press Field Guide

Rocky Mountain Natural History: Grand Teton to Jasper

IN COLLABORATION

National Audubon Society Field Guide to the Rocky Mountain States

National Audubon Society Field Guide to the Pacific Northwest

National Wildlife Federation Field Guide to Trees of North America

America from the Air: A Guide to the Landscape Along Your Route
with James S. Jackson

TREES IN TROUBLE

Wildfires, Infestations, and Climate Change

DANIEL MATHEWS

Illustrations by Matt Strieby

COUNTERPOINT
Berkeley, California

Trees in Trouble

Library of Congress Cataloging-in-Publication Data
Names: Mathews, Daniel, 1948– author.
Title: Trees in trouble : wildfires, infestations, and climate change / Daniel Mathews.
Description: Berkeley, California : Counterpoint Press, [2020] | Includes bibliographical references and index.
Identifiers: LCCN 2019017930 | ISBN 9781640091351 (alk. paper)
Subjects: LCSH: Forest health—West (U.S.) | Trees—West (U.S.) | Forests and forestry—West (U.S.) | Forest fires—West (U.S.)— Prevention and control. | Forest monitoring—West (U.S.)
Classification: LCC SB763.W38 M38 2020 | DDC 634.90978—dc23
LC record available at https://lccn.loc.gov/2019017930

Jacket design by Jenny Carrow
Book design by Jordan Koluch
Illustrations by Matt Strieby

COUNTERPOINT
2560 Ninth Street, Suite 318
Berkeley, CA 94710
www.counterpointpress.com

Printed in the United States of America
Distributed by Publishers Group West

10 9 8 7 6 5 4 3 2 1

This book is for the standing people,
that people walking among them
may know and honor them.

Contents

Introduction 3

1 A Loaded Atmosphere 9

2 Inferno. 29

3 Outbreak 59

4 Cookie Cutters 91

5 The Bleeding Edge 109

6 Thin and Burn 125

7 North, and Up 163

8 Ghosts 183

9 Fading White. 193

10 Resistance 209

11 The Enduring. 217

12 Future Forests 227

Afterword. 237

Acknowledgments 247

Notes 251

Index 281

Trees in Trouble

INTRODUCTION

Gary and Lew are sitting by the fire—deep in the Sierra.

Lew says, "You think rocks pay the trees any mind?"

"I don't know," says Gary. "What are you getting at?"

"Well," says Lew, "the trees—they're just passing through."

LEW WELCH AND GARY SNYDER, AS TOLD BY BILL YAKE

In western North America there are living pine trees older than the Egyptian pyramids. They survived through millennia of relatively stable climate. It's different now: in our lifetime we will see many trees just passing through. Groves where we sought refuge will themselves be refugees. Some tree species will disappear locally or regionally. Some forestland will turn into grassland, chaparral, sagebrush, or weedy wasteland. The forests of 2050, on a great many sites, will barely resemble the forests of 1950. Some of these transformations are already happening. In one-third of the low-elevation pine forest area in the U.S. Rocky Mountains that has burned in wildfire in this century, new conifer seedlings are failing to show up.

Climate change, in concert with pests, pathogens, and decades of misconceived fire suppression, is causing these sweeping changes. But there are actions we can take to limit the damage. This is a book for everyone who cares what happens to these trees, groves, and landscapes.

The book focuses on pines—two dozen species that stand as living emblems of the North American West. Growing in different elevation zones and employing utterly different adaptive strategies, they face a variety of threats today. In different ways, the threats are all affected by a warmer, effectively drier climate and a consequent resurgence of bigger, fiercer fire.

Seeing smoke, charcoal, and rust-colored vistas of the dying and dead, many people are upset. Few understand that human choices continue to play strong roles in the fate of these landscapes—and I'm not just talking about halting climate change. There are better ways to treat our forests and to live our lives around them. Public discussion of forest issues is dominated by simplifications (e.g., "the forests are all sick because they grew too dense" or "because they haven't been managed") that interest groups and politicians hijack and build into myths and blame games. Most of our forests have in fact been managed, and it hasn't made them less vulnerable to drought, insect pests, diseases, and fire.

Scores of forest scientists—many of whom I visit at field sites in the course of this book—are out there figuring out how the forests work, why they are not working so well today, and what management practices hold the most promise. There is a middle ground in between clear-cut logging and just shutting our eyes and hoping. Chainsaws can be tools of intelligent forestry.

Forest stewardship once held a place near the core of America's vision of itself. Tourism, skiing, and vacation homes all descend disproportionately on the West's pine country. Trees are essential to the image of most of our national parks. Three national parks were established to protect groves of the three tree species that rank as the world's biggest, tallest, and oldest.

But neither leaving lands untouched within national parks and wilderness areas nor developing boutique neighborhoods protects trees from climate chaos. The challenge is enormous; to meet it is crucial. The question is not whether forestry can restore the status quo of 1870 or even of 2000—it cannot—but how it can influence forests for the better. If we treat our forests right, we can at least ameliorate the declines in forest extent and diversity and the consequent impoverishment of the aesthetic, economic, climatic, and spiritual benefits we count on from them.

Like most of the forest ecologists I have met (and like me), Nate Stephenson initially got into this line of work because he loves mountains, forests, and backpacking. He loves backpacking so much that he doesn't even go away for his vacation every summer; he just goes backpacking right there in the national parks where he works—Sequoia and Kings Canyon. Like several other scientists hammered by shocking losses of trees in the terrain they know and love best, he doesn't waste time wallowing in the heartbreak. Much of his focus is on the amazement and scientific interest in finding himself in the middle of unexpected ecological transformation. "This is giving us hints of what we might see happen in the future, so we're just trying to learn as much as possible from it."

At the same time, Stephenson does want to send out an alarm right now:

More and more, what worries me is that the changes we're going to see are going to be traumatic and sudden. People have a tendency to think overall the temperature is rising slowly, so overall the changes in the ecosystems will be slow. I think we're getting better and better evidence—and pines are a prime example of that—that no, your changes might be extremely abrupt. Is there a way that we might ease that transition more? Rather than having it be really abrupt and suddenly there's no habitat left for a lot of forest-dependent creatures, and you get erosion, and you just lose a lot of what forests do for you. Can we anticipate that and actually start easing the transition ahead of time?

THE PINES

Threats to forests today are bewilderingly numerous. This book narrows the focus to the North American West and, further, to the pine trees, the genus *Pinus*. This simplifies the story while still broadly covering climate-related issues in the West. It gives us one major class of insect scourge, the bark beetles; one devastating exotic disease, white pine blister rust; and three very different fire regimes to which our pine species are adapted in very different ways. It leaves out the wet Northwest Coast, where the tallest trees and most massive forests grow. (Those forests are not yet suffering from climate change in ways that scientists can resolve sharply—though there are some ominous signs.) So we're talking about the parts of the West where pines grow. Sometimes called the dry forests, they include the very driest places that can grow forests in North America.

There are 113 species of pine worldwide—far more species than any other genus of conifer. They bestride other conifers even more colossally in terms of geographic range, or number of individuals, or board feet of lumber, or food uses (pine nuts and pine cambium), or nonfood nontimber economic uses (turpentine and pitch, which was once the indispensable caulking for ships), or just about any other metric that humans have shown an interest in. Except sheer size.

Pines are not known to pine away; etymologically, the verb *to pine* relates to the noun *pain*, not to the pine tree. A few hundred years ago, the English noun *pine* and its European-language counterparts were applied to conifers broadly: when David Douglas found the Douglas-fir tree he called it a pine. But over time, taxonomic concepts narrowed and vernacular usage, in this case, followed along. Today's pines separate themselves from other conifers by several characteristics, of which just one is easily visible and applies across the board: the long, slender needles grow in bunches wrapped together at their bases by small, brownish, papery bracts. Even the single-leaf piñon pine, which has de-evolved to where it typically has only one needle per "bunch," still has those bracts. No such bracts are found outside genus *Pinus*.

To identify species of pines, we can start by counting the needles in the bunch. Usually it's two, three, or five. While that can get you started on ID, some species vary in their needle bunches. To taxonomists, there are two subgeneric sections, *Pinus* and *Strobus*. Subgeneric distinctions don't matter just to taxonomists; they matter, for example, to white pine blister rust, a fungus that attacks only subgenus *Strobus*. Section *Pinus* includes most two-needle and three-needle species. *Strobus* includes the five-needle "white pines" and also the piñon pines, whose bunches range from a single needle on up to five or occasionally eight.

Pines feature the longest needles, the biggest and heaviest cones, the greatest production of pitch. While the long needles offer a lovely, soft visual effect and the pitch emits a nice sharp note to our noses, those characteristics aren't there to please us; they are crucial adaptive traits. Long, bunched needles and big cones serve to carry surface fire, which some pines rely on. Big cones carry big rich seeds to attract animals as seed dispersers. Pitch is not a pine's lifeblood (that would be sap) but its defensive weapon—the barrels of boiling oil it dumps from the parapets to engulf and poison bark beetles.

Ponderosa Pine

1 A LOADED ATMOSPHERE

The aroma hits when I open the car door: warm butterscotch, vanilla, but with an edge, less saccharine, more toast. "You smell them before you see them," says the man sometimes known as the forest guru. "I can identify several species by their aromatic phenols alone." It's September in ponderosa pine country, and for eight hot, dry weeks the smell has ripened.

Ponderosa pine is the tree of the West—the species equally beloved in Colorado, Alberta, Arizona, Oregon, Idaho, New Mexico, British Columbia, Nevada, Washington, Wyoming, Utah, even South Dakota in the Black Hills. It plays a major role in around two-thirds of the forested acreage in the West. It shares California with its near-twin Jeffrey pine, which is equally or more fragrant.

In youth its bark is rough and blackish—colloquially, it's a "bull pine" or a "black pine." When it grows up it will become a "yellowbelly." Sometime between age forty and a hundred and twenty, the outer bark of the rough ridges begins to exfoliate and reveal colorful jigsaw-puzzle flakes. Orange-brown is typical, but yellow-ochre is mixed in, and brick red, while on some sides or some sites the overall cast is lilac-taupe with a few flakes in maroon and ochre for contrast. As the color enriches with age, so does the fragrance.

Over the next four centuries (for those lucky trees that make it), black gives way to warm colors as the colorful ridges widen into broad plates while the blackish furrows remain slim. The plates' width is a clue to the tree's age: when they are ten to twelve inches wide, it's likely older than four hundred. Almost every ponderosa pine over the age of 150 has survived fires. After a fire passes through, bark low on the tree is charred black for a few years, but as time goes by little flakes of bark keep popping off until a fresh colorful surface is exposed all across the plate. Charred black bark persists in the furrows.

Scientists who study fire history by means of tree rings cut cross-sections out of stumps or fallen trees; they sand them to a high polish to make every tree ring visible, revealing a graphic image of all the fire scars the tree received, datable to the year and even the season when the scars formed. Studies of thousands of these "cookies" tell us that ponderosa pine communities typically experienced surface fires every few years. Picture these as grass and brush fires, neither tall nor particularly hot. They weeded out most conifer saplings and a few bigger trees; most important, they tended to weed out other tree species. Ponderosa and Jeffrey pines were disproportionately represented among the few trees surviving multiple low fires, thanks to substantially thicker bark at the sapling stage, and to buds that sometimes even enable them to survive complete scorching of their foliage. Here and there the flames of a low fire leapt up and torched an old pine, too, but enough survived to account for all the yellowbellies we see today.

Ponderosa pines don't just tolerate low-severity fires, they kindle them. Whereas most western conifers have small needles that fall to earth and decompose into a compact duff that holds more moisture, long pine needles dry out and persist on the ground, remaining highly flammable. Since they fall as three needles bundled together at one end, they also hang up on shrub twigs, becoming "needle drape." The dry needles provide exceptionally hot-burning fine fuel, and fallen pine cones provide the next bigger size of kindling. In consequence, ponderosa pine needles and cones can carry a low fire, with or without the help of understory plants.

Humans have long admired the style of forest that resulted: big, old, fragrant yellowbellies outnumbering the other trees, columnar, widely spaced, with a parklike understory you can easily walk through—or drive your team of oxen through, hauling your covered wagon over the Oregon Trail:

> Our drive of the forenoon of [September 8, 1853] was still among the pine openings. The atmosphere was loaded with balm.
>
> I can almost say I never saw anything more beautiful, . . . the forests so different from anything I have seen before. The country all through is burnt over, so often there is not the least underbrush, but the grass grows thick and beautiful. It is now ripe and yellow and in the spaces between the groves (which are large and many) looks like fields of grain ripened, ready for the harvest.

Fires prior to white settlement were lit sometimes by lightning, and sometimes by Indigenous people who used fire to facilitate hunting and travel, to wage war, to promote huckleberry crops, and for many other reasons. When the white men pushed the natives off the land, the low fires pretty much ceased.

Don't get me wrong: the white men loved fires, too, and started plenty of them—to clear land for farming, to see the rock formations better if they were miners, or by accident, or sometimes for the sheer joy of it. They had no magic to snuff out lightning fires, either. But they brought cattle and sheep. The rich grass in fire-groomed ponderosa pine communities lured the cattlemen, and later, after the cattle devoured the grass down to the nubs, the sheepmen. "The first thing we did," says the forest guru, "was graze the crap out of this country, from Arizona to Canada."

Without a continuous grassy understory to carry the fire, fires no longer curled around the feet of the ponderosa pines. Eventually Smokey Bear got involved in putting fires out, with varying success, but in this story it was the grazing that came first and changed the scene. Without frequent light fires, the wide spaces between big pines filled in with smaller trees. Whereas fires in the low understories rarely reached the limbs and foliage of big ponderosa pines, fires in these filled-in communities may do so; the infill trees are called "ladder fuels" because flames climb them from the grass and brush into the formerly safe pine crowns. Those low fires also stimulated ponderosa pines' natural defenses against bark beetles. Now the big old pines are threatened. On this clear September day, the forest guru is here to save them.

His name is Jerry Forest Franklin. His parents gave him the middle name Forest; much later, in the 1980s, journalists came up with "forest guru" after they started looking into this new quasi-mystical notion of "old-growth forests" and kept finding his fingerprints whenever they got near the bottom of the story.

In those days he was a Forest Service research scientist, an organizer and inspirer of a whole community of scientists. The guru role was implicit, at most. Today he is officially a professor, and he is here in south-central Oregon with students, twenty-five of them, several of them graduates who took the course at least once as undergraduates but keep coming back for seconds. The course, Ecosystem Management ESRM 425, consists of a two-week camping trip conducted before the University of Washington's fall term gets underway.

Franklin is a tall and vigorous seventy-seven-year-old in a plaid shirt and a much-crushed gray wide-brimmed fedora. Accompanying him are Norm Johnson, Franklin's colleague and coauthor of many years, in plaid shirt, much-crushed brown wide-brimmed fedora, and gray mustache; and Strider, one of a series of golden retrievers Franklin has had at his side over the decades. Slender, lively, friendly, rarely barks, doesn't waste time retrieving sticks, has his own carpeted space the full width of Franklin's dust-laden Toyota Tacoma.

The students have spent a day in a pine forest on U.S. Forest Service land in southern Oregon, marking trees with nylon tape to indicate which to cut and which to leave. They appraise the species, the size, the spatial patterns of the trees. Clumps are good; gaps are good; every old pine is untouchable; every midsize white fir next to an old pine is ladder fuel, take it out!

This is to be a "restoration harvest." The chief goal is to preserve the old-growth ponderosa pine ecosystem by restoring conditions in which the pines are more resistant to fire, bark beetles, and drought. Other objectives include biodiversity, supporting populations of deer and elk (particularly to benefit the Klamath Tribes, who have treaty rights to hunt here), and sending some wood to the sawmill when appropriate.

"The most important thing," to Franklin, "is that we respect the forest. My feeling is that if the forest were sentient, it wouldn't object to being seen as having utilitarian value. It would object to being treated as an object."

In the 1980s and '90s, Franklin's effective advocacy for old growth roused bitter opposition within logging communities, the timber industry, and much of Oregon's College of Forestry. One faculty member, William Atkinson, fumed to a reporter, "The Chinese have Chairman Mao; we've got Jerry." He was an environmentalist hero; but today, his continuing advocacy for old trees puts him at odds with many environmentalists who see all chainsaws as threats, and fires of all kinds as benign. It doesn't matter to him which side the flak is coming from. "I eat flak for breakfast," he replied in 2006 when Johnson warned him in advance of a stormy public hearing.

Extending the Mao metaphor, Johnson and Franklin are half of the "Gang of Four," a scientific team mandated by Congress in 1989, whose work led to President Clinton's Northwest Forest Plan. The moniker "Gang of Four" may have originated in the same industry circles that compared Franklin to Mao, but before long the four embraced it.

If we wanted to distill Franklin's career into a bumper sticker, it might read SAVE THE BIG OLD TREES. In the past twenty years he has focused much of his energies on the "dry forests" where pines rule, but the core years of his career were spent in the wet forests on the wet west side of the Cascade Range, and particularly on the H. J. Andrews Experimental Forest. That began with his summer job at age twenty-one. The Forest Service dropped him off for two nights of backpacking solo, to survey the watershed divide bounding the H. J. Andrews.

"I was so green! I didn't sleep at all for those two nights, because I was afraid of bears. After that, I decided that to be able to sleep alone in the mountains without being afraid I had to keep doing it all summer."

(I'm reminded of my own first morning of backpacking alone, age nineteen. I was at least as green but more foolish: clambering down from my hammock in the morning and seeing a black bear, I tiptoed barefoot to see how close to it I could stalk. Luck smiled on me, as the bear chose discretion over valor.)

Franklin wanted to be a forester from the age of nine. He grew up in Camas, Washington, a few miles up the Columbia from Portland. His father worked in the paper mill there for thirty years. The family had moved there from Waldport, on the Oregon Coast, where their small-town grocery gradually went under during the Depression. When they got their first car, a 1937 Plymouth, when Jerry was eight, they couldn't wait to take it camping in the deep green forests of the Cascades. That did it for Jerry.

The H. J. Andrews Forest was set up for experiments in how best to cut all the trees down and how best to grow the next crop of them. Franklin got to thinking about the scattered really old forests that remained. No one had even looked at them yet, almost literally. No one had gone into the ancient forest canopy, where hundreds of species awaited discovery and naming—insects, spiders, lichens, and one salamander. No one knew that the big fallen logs typically take six or seven centuries to rot away, during which time they serve as the seedbed several of the dominant conifer species require. Everyone thought that trees used their own roots to draw water and nutrients out of the soil, when in fact that function is performed mainly by fungal networks that also transfer materials from tree to tree, for nurture and, some argue, for communication. And no one knew, least of all the School of Forestry.

Franklin wondered whether converting all the Northwest's old growth to even-aged plantations—which was absolutely the plan in 1957, and the entire basis of his course of education in forestry—might lose some old-growth features that were actually valuable, economically as well as ecologically and scientifically. Within a decade, he and his colleagues utterly transformed the H. J. Andrews Forest. It became the leading site for the study of natural Northwest forests.

The forest ecology and management issues that Franklin and Johnson addressed throughout their careers were already insanely complex and surprisingly little studied. (Humans, who snicker at the idea of watching paint dry, have a hard time watching processes unfold at a forest's pace.) There was competition, forest succession, pests, diseases, invasive species, economics, exurban development, and fire. Considering all those disparate factors, let's call that a three-dimensional problem. Well, in the last twenty years it's as if we've discovered that forests live and die in four dimensions, or really that each of those complicating factors has a fourth dimension: climate change. As its practitioners like to quip, forest ecology "isn't rocket science. No, it's much more complicated than that."

Today's climate crisis changes everything. No matter whether you're talking about trees being killed by fires or by insects or by mystery syndromes like Sudden Aspen Decline, climate is there, putting its big thumb on the scales of life or death. (That's why this book has no single chapter focused on climate change effects.) Even though we don't know which parts of the West will be getting more or less precipitation, we know that they are all getting warmer, and warming at any given moisture level increases drought stress in trees. The force that sucks water out of a plant into the air is captured by a measure called vapor pressure deficit, which is simply temperature times relative humidity. It always goes up when temperature goes up. Drought stress increases a tree's likelihood of dying in any given fire, or when attacked by insects. And it can also kill them directly. (A large increase in precipitation, if it were effective during most of the growing season, could theoretically save trees from drought stress, but that is not the sort of precipitation increase that anyone foresees in this region.)

Climate has already warmed since pioneer days. It has warmed more in the American West—two to four degrees Fahrenheit in a hundred years—than in most parts of the continent other than the far North.

Perhaps forest "restoration" is misleadingly named. Franklin insists that the point is to restore not a past condition, but a more resilient condition.

Since the forests lived for the last five hundred years in a climate that does not exist today and certainly won't prevail for the lifetimes of today's seedlings, we have to try to figure out what will work best. Scientists study forest conditions back in 1853 because that's their window on natural forest processes that operate on too slow a scale to study in a field experiment—processes the ponderosa pine is exquisitely adapted for.

Some enviro activists would apparently rather take their chances with forest fire than with chainsaws, even chainsaws for restoration thinning. They doubt that we know enough about how forests work to be able to take over and do a better job than nature will.

But nature's loving care is no longer an available option because of the powerful sweep of human influence, including abnormally abrupt climate change. Franklin tells his students, "this forest is not going to sort itself out. People say, 'let nature take charge.' Nature will take charge of it, but not in a way that we will like. We will lose resources, lose old trees, which are essentially irreplaceable. People who say let nature take care of itself are being totally irresponsible. We are responsible for how this forest is today."

He never tires of telling about the day he backed an activist up against a magnificent old ponderosa and said, "This tree is gonna DIE, and it's gonna be YOUR FAULT!"

Here's the logic: many old ponderosa pines will burn and die unless the ladder fuel trees around them are removed; commonly, the ladder fuel trees are already too big to be removed by fire without losing a lot of the big pines.

Hugh Safford, the Forest Service's top ecologist for the California region, paints a similar picture as he shows me around Emerald Point, the finest stand of trees in state park land on the shore of Lake Tahoe.

"The Tahoe Basin just got slaughtered in the 1870s."

The 1859 discovery of the Comstock Lode triggered the slaughter of Lake Tahoe's forests. Over the preceding decade, gold miners overran the central Sierra Nevada using technology ranging from panning for gold on up through gravel-sorting troughs and rockers. All these technologies consumed lumber, but not at the rate that was soon to come.

The Comstock Lode, in Nevada around Virginia City, was a buried silver deposit with minor gold deposits in it here and there. It required digging tens of feet down through crumbly ores highly prone to collapse. To keep the mines open and the miners alive for a while, massive timbers were emplaced underground in "square-set" modular frameworks. As the chronicler of the day described it, the Comstock Lode became "the tomb of the forest of the Sierra." While 750 million board feet of the best lumber was buried in mines, easily five times that volume was burned to fire the industry's steam engines and smelters. Moving all this wood from the Sierra to Nevada took still more wood: three major fluming companies built short railway or tramway lines to get lumber from mills on the lakeshore up to passes where it could be transferred to wooden flumes on high wooden trestles for the downhill part of its trip.

Supplying this timber was lucrative enough that a new arrival named Sam Clemens attempted to make his fortune by claiming a three-hundred-acre "timber ranch" on the lakeshore. As he spun the tale in *Roughing It*—written under a new pen name, Mark Twain—he didn't last long as a lumberman, as he soon incinerated his chaparral: "Looking up I saw that my [cooking] fire was galloping all over the premises!"

Twain's chapter on his Lake Tahoe summer, leading up to the fire, reads as an idyll of grown men inventing summer camp for their own pleasure. While he and his friend were lolling around on a zero budget, the tycoon Ben Holladay bought lakefront land and, in 1863, built himself a summer lodge near Emerald Bay. Holladay's wealth came from his stagecoach line; he would go on to buy the Pony Express and build the Oregon and California Railroad. Another man bought five hundred acres on Emerald Bay for a resort. By 1898, when the Comstock Lode mines closed, most good forestland in the lower parts of the Tahoe basin was stripped bare, but Emerald Bay stood out, the jewel of the lakeshore, saved by these plutocrats who preserved their trees for their scenic value.

But saving them from the saw did not protect them from what the next century would bring—firefighters.

That's why Hugh Safford wants to show me Emerald Point today. "We have a photo of it from 1886. You can recognize the same big trees that are here today, but in 1886 they had space between them. Now they're completely filled in with trees of all sizes. One cigarette gets tossed on a really dry, windy day, and this is gone. Every year they put out two or three ignitions in the area. That's what I lose sleep over."

I think he's being literal about losing sleep; I've heard the anguish in his voice as he's brought up Emerald Point again and again.

It's easy to see why density makes forests fire-prone: the trees are close enough together for flame to carry from one tree crown to the next. In classic Jeffrey pine forests maintained by frequent fire, most are too far apart for that. Additionally, the lowest limbs of the taller trees are too high for the flames of grasses or shrubs to reach. If a tree sapling happens to make it to, say, twenty feet tall, a grass or duff fire may or may not cook it to death, but the odds of its crown catching fire are lower than when it was smaller, and if that happens, it may or may not scorch or ignite a bigger tree's crown above it; and even if that happens, it probably won't jump to the next tree crown over. A few canopy trees may die without catching fire, if low flames overheat their roots or the moist cambium behind the bark; a few may die during the years following the fire, where scorched bark draws a bark beetle attack. When you add up all the mortality, it's still just a small percentage of the canopy trees. That low percentage makes it, by definition, a low-severity fire.

Safford is a lean six feet tall, with a frequent grin, blue eyes, and Nordic bone structure. He grew up on a Montana ranch that subsequently succumbed to Bozeman sprawl. (At age seventeen I was advised that everyone from Montana is really nice; it hasn't failed me yet.) He got a degree in geology, taught high school science for a few years (three of them in Brazil), and then began looking for work on the ecology side of things. On the surface he's fast-talking, fun-loving, California outdoorsy. Appearances belie a ferocious overachiever. He goes rock climbing three times a week and mountain biking most of the other days. Meanwhile he holds down two jobs—region ecologist for the Forest Service, plus research faculty at UC Davis. In the latter capacity he works closely with a succession of grad students and postdocs on their research projects, for which he also writes the grant applications. All summer, young researchers (and Safford's son Marcel, working for the Forest Service the summer when I visited) come and go from the couches and beds in his small house near Lake Tahoe. With Kelly, his wife of twenty years, who also works, he has raised two sons. His name is on six or eight published papers a year. He speaks six languages fluently, holds a permanent resident card in Brazil, and provides advice to resource and fire managers in the tropics, the Middle East, Mexico, Spain, and Portugal, spending months at a time working abroad. His cosmopolitan perspective slips out as acerbic comments on ways the United States hasn't caught up with the rest of the world: the inadequacy of our mass transit; how best to catch speeders; or the advantages of mountain highways going through tunnels rather than crossing rockslides.

Like Jerry Franklin, Safford has forestry in his middle name, Deforest. His Swedish grandfather had trouble pronouncing "Hugh," so he called him the Swedish word for "forest." (And then there's the coincidence of his last name: my grandmother was a Safford, and there aren't all that many Saffords in the country. Who knows?)

At Emerald Point, the giant trees in the old stand are incense cedars, uncommonly fat ponderosa pines, and ponderosa's close relative, the Jeffrey pine. It's named after a mysterious early plant hunter, John Jeffrey, who disappeared at the height of the gold rush, perhaps finding more allure in the prospect of gold than of pressed plants. We measure a ponderosa seven feet in diameter, a cedar more than eight feet. Using an elegantly simple tool, an increment borer, we auger a four-millimeter core out of a young cedar to check its age, and yes, it began life in the 1890s, like most of this stand's infill. I sniff the core, ahhhhh . . . the tang of childhood's pencil sharpeners. Incense cedar is the archetypal pencil wood. Ticonderogas, for one.

The sharp fragrance contrasts with Emerald Point's ambient smell, which resembles the balm of ponderosa pines, but which in the Tahoe area emanates instead from Jeffrey pines. I put my nose into crevices of both pine species growing next to each other, and the Jeffrey pine phenols smell stronger and sweeter than ponderosa here. (To the forest guru they might be as distinct as Obsession from Chanel No. 5; but I swear they aren't different enough to warrant the weird claim of some California botanists that they identify ponderosa pine by the *absence* of any vanilla fragrance.)

At a breakfast café I conduct a brief survey on vernacular awareness of Tahoe's amazing aroma. Yes, when they return from Vegas or the Bay Area they thrill to the vanilla scent of home. No, they didn't know that it's the smell of a particular species of tree, the very one that shades the café parking lot.

Jeffrey pine occupies a range whose bounds match those of California's Mediterranean climate, characterized by at least five very dry summer months alternating with rather wet winters. Ponderosa pine co-occupies much of that range, but only on the more benign sites, whereas Jeffrey pine reaches both dryer and colder parts of the Sierra Nevada, and also grows readily in serpentine soils. Serpentine—California's state rock—weathers into nutrient-poor, metal-saturated soils that at least half of most regions' plant species can't tolerate. As a result, serpentine areas are conspicuous even from miles away because their plant cover differs in its species makeup and is altogether sparse. In the Klamath Mountains along the California/Oregon line, the climate is benign enough for ponderosa pine and Douglas-fir, which outcompete Jeffrey pine there, excluding it everywhere except on serpentine. Around Lake Tahoe, in contrast, Jeffrey pine rules; the dry summers must be a bit too long or too hot for ponderosa pine, which is uncommon and confined to moist, sheltered sites, like Emerald Point.

The two species are so closely related that they occasionally hybridize in the wild, and they can be hard to tell apart until you find cones: Jeffrey pine cones range to a much larger size, and have their own distinct shape of prickle at the end of each cone scale.

Hugh Safford is sorry he can't take me to see "pine openings" just like the ones on the Oregon Trail, not anywhere in the United States or Canada anyway. There are groves that come pretty close, sprinkled all over the West, but they've all been touched by fire exclusion; we can't be sure they've had all the fires they would have had if the government hadn't been putting most fires out. Most of them are overcrowded, or else were at some point, and were brought back with prescribed fires or thinning. Most lost a lot of their biggest pines to logging, perhaps so long ago that we may not know how many or how big.

But Safford does know a secret mountain range covered with pines that have seen only minimal fire suppression or logging. It sees few tourists but has become a mecca for dry-forest ecologists. Called Sierra San Pedro Mártir, it hosts a national park in the Mexican state of Baja California Norte, not far southeast of Tijuana. Jeffrey pines predominate, and the rest of the tree species are also shared with the Sierra Nevada. Chaparral covers steeper slopes. The pine forests are at around seven thousand feet, in a Mediterranean climate pretty similar to Lake Tahoe's shores, which are slightly lower. Most of the annual precipitation falls in winter as snow. Night skies glitter brilliantly in thin, dry air forty miles from the power grid.

It does have a history of grazing, but for whatever reason that did not put a crimp in the frequent fire regime here. A new policy of suppressing fires was put in place thirty years ago, but that's only now starting to show some signs of the dreaded effects. (Safford is working on getting them to stop the practice.) Enough fires still come through to demonstrate a preponderance of low-severity burns. Most remarkably, even in the fourth year of a four-year extreme drought topped off with bark beetles—the same combination that looks so scary in the Sierra Nevada—a fire burned through the pines and left 83 percent of them alive.

Safford's colleague Scott Stephens gave a conference talk that crescendoed on the words, "This gives me hope! When I see the Sierra San Pedro Mártir ecosystem, which is being thrown around by climate change, it burned in a wildfire after a severe four-year drought, and by the way, eighty-three percent of those trees survive, this tells me that there's hope . . . that you can actually have a forest that's kicked around and can kick back!"

A few blocks from the breakfast café, Safford shows me an anomalously treeless neighborhood of new homes with metal roofs and plastic decks.

These houses' predecessors along with their trees were vaporized by the Angora fire of 2007. Angora was not an especially large fire, but on a June afternoon of strong down-valley winds it took less than three hours to expand from an escaped campfire to a 250-home firestorm. No residents died, but few had more than ten minutes to collect their keepsakes.

The state already had guidelines for defensible (fire-inhibiting) space around homes, but these were barely enforced. Homeowners tend to push back hard against meddlesome and expensive government regulations—*until* something happens to convince them that forest fires actually do burn down houses right in their neck of the woods. Angora proved persuasive in that regard. (New building codes were enacted, but only for new construction, so they were of little benefit to Santa Rosa, Redding, or Paradise, where so many homes burned in 2017 and 2018.)

Angora provided a natural experiment where Safford could study the effects of fuel reduction treatments a mile from his home. The fuel treatments (thinning and prescribed burning) aced the test, showing up after the fire as sharp-edged patches of less intense burning. "The line is just stark, it's astounding." In this case, forest restoration treatments effectively limited fire intensity. Where the fire swept into treated patches, it dropped down to a low level despite the dry, windy conditions. But that didn't save the houses, which were poorly designed for resisting fire. Small embers lofted by wind were enough to ignite some houses, which soon ignited their neighbors, and then the surrounding forest. "The houses burned the forest," Safford tells me. Similarly in Santa Rosa, and in Waldo Canyon, Colorado, conflagration leapt from house to house with little help from trees.

Half an hour south and just over the top of the Sierra Nevada onto the west slope, Safford shows me a recent "good" burn, the Long fire. Sparked by lightning during fairly calm weather, "it was the kind of thing that starts and then just skunks around." Well removed from towns or summer cabins, and in a less flammable forest type that wasn't likely to blow up, it was allowed to burn rather than being fought immediately. (Eventually it expanded to where it could threaten structures, and firefighters were then sent in to contain it.) A good majority of the large canopy trees survived, and when we walk around in it five years later it is a fine, gorgeous stand, its red fir trunks cloaked in wolf lichens incandescing neon-yellow where the sun catches them.

Just half an hour farther west lies a counterexample, a crazy monster wildfire. In one day, September 17, 2014, the King fire spread by sixty-eight square miles. Picture a fire sweeping all of Manhattan in a day. Manhattan is thirty-four square miles, so now picture a double Manhattan. Most of those sixty-eight square miles, including one contiguous "patch" bigger than San Francisco, burned at high severity—a classification defined by at least 75 percent mortality of trees. (Intensity is a measure of a fire's energy, or its heat and flame height, whereas severity traditionally represents the percentage of overstory trees that died, either as individuals or as net basal area lost, so that ten big trees count more than ten small trees. Damage to the soil, or soil burn severity, is a third distinct measure. If you want to get literal, scientists still say severity represents a mortality percentage, but the actual severity figures they use nowadays usually come from analysis of spectral reflectance in satellite imagery; computers measure the degree of color change, pixel by pixel. There's ongoing debate over whether other approaches might be more realistic.) While 75 percent satisfies the definition, in fact every single tree died on many of those acres.

The King fire was ignited by Wayne Allen Huntsman, a four-time convicted felon chalking up a fifth, arson, which got him a twenty-year sentence and a restitution fine of $60 million, at least on paper. After he lit several fires a few feet apart, he shot a drunken-looking video of himself smirking, "I got fire all around me. Look at me, babe. I got fire right there! I'm stuck in the middle, babe!" We have no record of whether Babe was impressed. (Wayne's older sister, Tami Joy Huntsman, made the news with quotes about how he was a nice guy and she didn't believe he did this. Tami Joy, in a grisly turn of events, was arrested a year later for the torture and murder of a niece and nephews in her care, and eventually sentenced to five consecutive life terms plus nine more years.)

Another term for high-severity fire is "stand-replacing"; but when a high-severity patch extends for miles in every direction, it tends to get stuck at "stand-removing." Conifers generally reseed burns with seeds that fall or are windblown from trees still standing. (A few pine species enlist birds to cast their seeds much farther, but that's an exception.) Most burns retain at least a few such seed trees—either scattered survivors or trees killed by scorching of their trunks and roots, leaving many cones intact and full of seeds. But intense crown fires consume the cones along with the needles and fine branches. Where that happens over large contiguous areas, natural reseeding can take a century or longer; conifer seedlings encroach generation by generation from the edges inward. That process disfavors ponderosa and Jeffrey pines and favors trees with lightweight winged seeds that travel farther on the wind. As long as these three-needle pines can sustain their low-severity fire regime, they provide the seeds to maintain their own dominance; but where today's conditions allow large patches of high-severity crown fire to burn, the pine forest is likely to give way to brush or, with luck, gradually to a new forest with far fewer pines than before.

To ensure and accelerate growth of a new forest, the Forest Service will replant parts of the King fire area that are too far from surviving trees. A burn is an opportunity to replant with seed stock from farther south, which will be better adapted to the climate the trees will live in. Fortunately, Forest Service policies have been changing and now allow that to happen to a small degree, in place of a rigid insistence on local seed. Tree populations are incapable of migrating northward fast enough to keep up with the pace of twenty-first-century climate change. (Forests chased abrupt warmings during and terminating the ice ages, possibly about as abrupt as today's, but those were not transitions we would have liked living through. To put that another way, those novel, rapidly readjusting forest communities may not provide ecosystem services at the level that nine billion humans require. See chapter 7.) In addition, replanting needs to reflect up-to-date science: the future forest needs to be far more sparse, clumpy, patchy, and diverse than the one that the King fire devoured. Much of that one was plantation designed to maximize timber production—a hope that the King fire foiled, in spades.

In parts of the West, stand-replacing fires—including some big ones—were the norm prior to Anglo-American influence. And some western pines, like the lodgepole, are well adapted to such fire regimes, with specialized ways of reseeding large burns quickly. But high-severity fire played only a small role wherever ponderosa and Jeffrey pines are common. That makes up a hefty portion of forests in each of the contiguous western states. Of all the U.S. fires since 2000 where more than $20 million was spent on fire-fighting, 62 percent held at least some ponderosa pines. Within California, Safford tells me, this mixed-conifer forest (with those pines in the mix) "is where all the recreation happens, it's where the key animals are that we're sued on and that we're managing for—spotted owl, goshawk, fisher—it's where all the logging went on, where most of the mining happened, it's the water catchment for the whole state of California, and it's the landscape in which we actually do active management [i.e., logging or thinning]." He calculates that 4 to 8 percent of fire area within the mixed-conifer forest burned at stand-replacing severity during presettlement times, whereas since 1984 that percentage has climbed above 30. For Oregon and Washington's ponderosa forests, it was 6 to 9 percent high severity before settlement, jumping to 36 percent recently.

The King fire illustrates the proposition that fighting fires leads to worse fires: rather open forests that used to burn frequently at low severity had gradually changed, due to fire exclusion in combination with logging and some replanting over a century or more, into dense forests with a different mix of tree species that can more easily burn at high severity. And that's what they did. (The term "fire exclusion" combines fire suppression with other anthropogenic forces that have made fires rarer, such as grazing and ending Indigenous people's fire setting.)

Even so, under milder conditions contemporary forests can burn more benignly, as the Long fire did—mixed severity with high severity occurring in small patches. But that doesn't happen extensively enough, because under milder conditions we can, and usually do, put them out while they are still small. There's a terrible perverse consequence to putting out the fires in mild conditions in combination with not putting out (because we can't) the fires in bad conditions. Over time, those numerous extinguished fires don't add up to much of the total acreage affected by fire; they're overwhelmed statistically by the few fires that happen to catch a spell of really terrible fire weather and blow up, impervious to firefighting efforts. Those explosive days produce most of the burned acreage in recent years, and it's largely for that reason that high-severity patches have shot up as a proportion of the whole. And that isn't just a statistical blip: it's truly the trajectory we're on, one that hurtles toward a West in which most ponderosa or Jeffrey pine land will burn at high severity within a few decades, and big old pines will become very rare.

Does that mean that the trend toward worse fires would be alleviated, even now, if we simply stopped putting fires out? Yes, it could. This is "the fire suppression paradox." As Sean Parks and colleagues put it, "aggressive fire suppression reduces the occurrence of low severity fire, thereby increasing fuel on the landscape and selecting for higher severity fire when the inevitable fire occurs."

Realistically, in many areas, especially in California, the forests are now so homogenized, dense, and laden with dead trees that the "alleviation" of bad fires would be very slight. We need thinning and prescribed fire first to reduce the fuels.

But forest ecologists generally support a policy of letting a lot of the lightning-ignited fires burn, at least in the long run, and for some areas right away. That policy was adopted decades ago by at least four national parks—Yosemite, Kings Canyon, Sequoia, and Zion—and it has worked out well. The Forest Service has adopted similar policies for their own lands; yet aside from some wilderness areas (see chapter 6) these are very rarely followed in practice. Why not? Legal, social, and political forces are stacked against allowing fire. That needs to change.

Lodgepole Pine

2 INFERNO

In June 2011, following a very dry winter, the forest was exceptionally dry all across the Southwest. Up at 8,400 feet above sea level in New Mexico's Valles Caldera, wind funneled through a low saddle holding a few scattered homes comprising the rural community of Las Conchas. An aging aspen tree blew over, taking down a two-strand power line serving a few homes. The power lines sparked and ignited some grass, and the wind whipped the flames. The homeowners saw fire and called the first responders, who arrived soon after, just in time for a big *whooomph!* as the wind-whipped fire exploded into the tinder-dry ponderosa pine canopy; they backed away and called for a bigger response team. But this fire was growing so fast that no human effort would slow it down that day, not the tiniest bit. Over the next fourteen hours it torched another acre every second, incinerating an area larger than Washington, D.C.

Residents of Santa Fe got out their tripods and began filming a towering convection plume of smoke and vapor. At night they captured flames five hundred feet high from twenty-five miles away. In their time-lapse movies (which you can watch online) the monstrous smoking convection column rotates like a slow tornado and climbs to thirty thousand feet, where the jet stream pulls it eastward.

Fires typically calm down somewhat at night, simply because the ambient air cools down. Typically, they also move faster upslope than downslope. This one, astonishingly, blew up into a firestorm at 2:00 a.m. and tumbled *down*slope across several mesas. The one witness at that hour reported giant orange rolling barrels of flame enveloping more than two miles of mesa top.

The guilty aspen tree itself never caught fire, nor did the closest houses. Las Conchas soon became best known as the name of this fire. (It means "seashells"; no marine fossils would exist in this volcanic terrain, but shells may have come here through intertribal trade.) The wind blew the fire away from the aspen and the houses, toward Bandelier National Monument and Cochiti Pueblo, and into areas that had burned badly in the 1996 Dome fire and the 2000 Cerro Grande fire. Those twice-burned areas were initially the heart of darkness, a moonscape. After seven years of scant and weedy revegetation, they are now Exhibit A in climate change–related deforestation.

A forest fire hardly ever consumes the trees it kills. They have too much mass and too much residual moisture. If it's a crown fire, it incinerates the needles but leaves the trunk, the branches, and even a lot of the fine twigs. Many lethal fires don't even do that much: fire in the shrub-and-sapling layer can scorch the overstory trees' bark and do lethal damage to the water-conducting tissues under the bark. Even fires in the duff layer, or in the mosses and lichens that drape coastal rain forest trees, can sometimes kill big trees by cooking their roots or their cambium. Then the needles turn brown and stay in place on the tree for many months. In consequence, most high-mortality-rate burns look like thick forests of dead trees, either gray or red-brown in the first year, with blackened trunks. Over the next several years the bark sloughs off, and trees may be white for years before falling.

No, to get a burnt-over moonscape you typically need two intense forest fires ten to twenty years apart. It needs to be a dry year in a dry climate. After ten-plus years the standing dead have largely fallen and dried out so that they can now be consumed by fire, and brush or conifer saplings (new plantations of them, in some cases) have grown in densely and can be consumed in the same flames that take the very dry old logs. Even then, it isn't exactly normal for the larger logs and standing dead snags to vanish from the face of the earth. It helps to have an extraordinary conflagration—a phenomenon seen mainly in German and Japanese cities bombed in the Second World War, and variously described as a firestorm, a "fire that makes its own weather," or a mass fire. This seems to have happened at Las Conchas, developing in the mixed-conifer forests and sweeping into the brush fields and dry fallen snags of earlier burns. After months of drought, brush, snags, and logs burn really hot, much hotter than the live trees, which necessarily have been sucking up and storing whatever moisture is available.

There's no way to prepare yourself for the scene of utter devastation at Las Conchas. (Or at Colorado's Hayman fire, or Arizona's Rodeo-Chediski.) It's a staggering, numbing experience. Three facts strike home:

1. Sheer scale. With no trees you see miles in every direction, and it's all the same, almost a desert. Splintery white snags of old trees, but even those are few and sparse. Decrepit clumps of green trees are far away and skimpy—like the post–meteorite impact world constructed for *The Road*.
2. This was a pine forest, a fragrant, verdant mountain slope, for centuries with only small-scale temporary lapses.
3. It won't be pine forest again. Not in our lifetimes, probably not in any foreseeable time frame.

There are almost no conifer seedlings. There isn't even much native southwestern brush growing. The soil looks like barren desert bearing scattered scruffy, mostly weedy plants, tumbleweed. A southwestern Gambel oak brush field developing seems almost the best we can hope for. So far, thorny New Mexico locust outnumbers the oak.

My Virgil guiding me into this hell, Craig Allen, may be as close as I've found to a media star among forest ecologists. The media call on him when they want to hear what's going on with southwestern forest fires. While he acts as a Virgil for the reporter, to the scientific world he sounds more like a Cassandra, publishing titles like "Predictions of Massive Conifer Mortality Due to Chronic Temperature Rise," "Regional Vegetation Die-Off in Response to Global-Change-Type Drought," "Climate-Induced Forest Dieback: An Escalating Global Phenomenon?," and "Global Vulnerability to Tree Mortality and Forest Die-Off from Hotter Drought in the Anthropocene." (A paper he titled in a relatively muted tone has been cited 4,202 times, which I think may be the most ever for a scientific article on forest ecology.)

But in person he is not the doomy prophet I expected. I'm welcomed by smiling eyes behind rimless glasses, an unquenchable stream of science talk, a trim sixty-year-old with straight salt-and-pepper hair emerging from under a beat-up Tilley hat. He invites me into his comfortable Santa Fe–style home in Santa Fe, where my eyes go straight to the collection of four giant pine cones: Coulter pine, gray pine, Jeffrey pine, sugar pine.

It's a great start at bonding, since I have the exact same four at my home. He wants to give me all the background info using his past presentations, on his laptop, before we head out into the field. For efficiency's sake. Two and a half hours later, his talk gradually accelerates in trying to squeeze everything in, repeatedly punctuated with "We*any*way . . ." as he realizes it's high time to head out.

Allen grew up the oldest of six kids, enacting Daniel Boone fantasies on forty acres of cutover Wisconsin woods his grandfather took on as a project. Dad taught high school biology. In graduate school Allen adopted the Jemez Mountains of New Mexico as his home terrain—rich green forests of a different style altogether. He studied them for fifteen years before they gave him their first incendiary clue that he might outlive them—the Dome fire.

After another fifteen years, Las Conchas.

"Two months post-fire, we walked out there, I saw two birds the entire day, a peregrine falcon in the distance, and a swift. There was no base to the food web! I saw one or two harvester ant mounds, they were inactive . . . there were no insects! There was nothing! It was deathly. This was one of the most severely burned places I have ever seen, because it just . . ."

He trails off. For once, words fail him.

Perfect: peregrine means wanderer, swift means . . . yeah, you get it: two birds known for roaming far and wide in search of morsels. Their passage is no indication of sustenance nearby.

Allen had to pack up and evacuate his office in Bandelier National Monument for the Cerro Grande fire of 2000 and again for Las Conchas in 2011. As soon as the threat of Bandelier's visitor center burning up had passed, at least for 2011, he was warning staff that they needed to spring into action to save it from washing away in flash floods. New Mexico gets serious rains (monsoons, technically) in July or August, and the incineration of Frijoles Canyon's plant cover was bound to have cataclysmic effects on runoff. Plant cover intercepts raindrops, breaking their fall and leading more of the water to percolate into the soil. After a severe fire, rainstorms instantly turn into surface runoff. Allen advised the monument's supervisor to destroy a small road bridge over the creek to prevent trees from stacking up against it in a flood, creating a dam that would raise the water even higher. This was done, and did in fact just barely save the visitor center. The flow rate during the flood was a hundred times that of flash floods measured there before the severe fires in this watershed.

That was just a minor facet of the year's post-fire flood problems. That same June, the week before the Las Conchas fire blackened mountain slopes west of Santa Fe, the mountains rising from Santa Fe's east side had their own fire with twenty-five square miles of high mortality. In August, sediments flooded down from the Las Conchas burn, rendering water intakes from the Rio Grande unusable for several weeks, for both Santa Fe and Albuquerque. New Mexicans realized they could no longer count on a safe water supply.

Public opinion of forest fires was transformed by the realization that even if fire never reaches their homes directly, its effects on water can still existentially threaten human occupancy of the region. Forest ecologists who had until then struggled to spread the gospel of forest restoration suddenly found a rapt audience. New Mexico today largely supports forest restoration as a way to reduce the wildfire risk. They call it watershed restoration. This level of popular support for prescribed burning is almost unique in the West. (It's rivaled by Flagstaff, Arizona, which voted in a major bond to fund forest restoration.) In Santa Fe it's especially remarkable considering that it's next door to Los Alamos, where the Cerro Grande fire destroyed 235 houses in America's most notorious and costly case of an escaped prescribed fire.

(In the Southeast, in contrast, Floridians and Georgians have been burning their longleaf pine forest for the past fifty years with little local opposition beyond some griping about the smoke.)

While floods and mudslides are the chief concern affecting water supply, there are also direct effects on water supply. Fires can increase the total amount of water coming down from the mountains over a longer term. This effect derives from thinning the forest, regardless of whether that's done by severe wildfires, prescribed fires, thinning with chainsaws, or even clearcuts. A forest is a massive pump sucking water out of the soil and transpiring it into the air. When you sharply curtail tree density, you shrink that pump and leave more water to flow into streams and rivers. A California study estimates that thinning forests to presettlement density all across a major river watershed would put 5 to 10 percent more water in the river. The effect became perfectly visible in parts of Yosemite National Park: a big creek drainage grew dense from 1890 until 1972, at which point the park started letting fires burn again. Since then, three-quarters of the basin has burned at least once, reducing density close to nineteenth-century levels. In several spots where forests with dry soils had encroached into wet meadows, fires converted these sites back to soggy meadows. Neither rain nor snow amounts can account for these near-wetlands' death and resurrection; it was caused purely by the number of trees.

But in parts of the Southwest, a surprising opposite effect showed up: when piñon-juniper woodland was sharply thinned by bark beetles, stream-flow was diminished rather than augmented. The researchers could only speculate as to the mechanism: presumably the decrease in transpiration from piñon needles was outweighed by increases in evaporation from the soil surface and the snow surface when they became more exposed to sun and wind. A West-wide review of studies on the hydrologic effects of fire found a mixed bag.

Since Las Conchas threatened the Los Alamos National Laboratory, it attracted the attention of some very smart physicists with access to very powerful computers, yet a conclusive analysis of the physics involved has yet to appear. Wildfire dynamics remain poorly understood, and that's especially true of mass fire dynamics. Until very recently, no one ever got instruments close enough to big wildfires to measure them.

Craig Clements, of San Jose State University, refuses to settle for this "observational deficit." His solution is RaDFIRE, the Rapid Deployments to Wildfires Experiment, a "meteorological field campaign aimed at observing fire–atmosphere interactions during active wildfire." The RaDFIRE team can chase firestorms. Everyone on the team goes through annual firefighter training and carries a red card to prove it. And they have technology never seen elsewhere in ground-based wildfire study. Their first Ford F250 crew-cab pickup truck has scanning Doppler lidar mounted to its bed, along with an automated weather station and a microwave profiler. Their second has mobile Doppler radar, which, he told me, "can see through any plume, and scan from anywhere. It's just awesome. With lidar we have to get pretty close." They can also release weather balloons to record radiosonde data. (Despite these unmatched capabilities, their funding grants have been hit-or-miss in recent years.)

RaDFIRE's work is still in early stages. In the first four years of operation they went to twenty-one California fires, including the King, Rim, and

Soberanes, and they flew over Pioneer in Idaho in an instrument-crammed plane borrowed from the University of Wyoming. "It's kind of like fishing," says Clements, a fly fisherman. "You go out hoping to find something."

Flying above the Pioneer fire, they measured the updraft in the convection plume at an off-the-scale 130 miles per hour, much stronger than anyone expected at that height. Downdrafts about half as strong (merely gale force) flanked the updraft on either side.

The team learned a lot about how fire convection plumes grow, intensify, and occasionally punch through the resistant tropopause layers in the atmosphere. The tropopause is the atmospheric boundary layer that acts as a kind of lid on moisture, clouds, weather, and updrafts. Over midlatitude regions it averages about thirty-three thousand feet up. If that number sounds familiar, that's because it's a typical cruising altitude of jetliners: planes climb to the tropopause to get above the weather, and to fly with less wind resistance, or with the help of the jet stream if they happen to be going the jet stream's way. As thunderstorm clouds develop, they get taller and taller, but if they get high enough, the tropopause forces them to halt and spread out, forming the familiar anvil-top shape.

Convection plumes rising from big fires are similar: they consist of hot air rising, and they're full of moisture, a by-product of combustion. Often they form puffy white clouds at their tops, called pyrocumulus clouds. They can create lightning up there. Only especially powerful ones called pyrocumulonimbus can punch through the tropopause into the stratosphere. Until around 2000 it was thought that only volcanoes were hot enough to punch through the tropopause and pollute the stratosphere with aerosol particles. Now, with the aid of satellites, we see that pyrocumulonimbus events are doing that in significant amounts almost every year.

The RaDFIRE team caught one pyrocumulonimbus forming, and they caught a decent vertical-axis vortex at one fire. It lasted half an hour and was ten times bigger than a fire whirl previously measured at a prescribed fire.

Still bigger and more powerful, by far, was the vortex on the outskirts of Redding, California, on the afternoon of July 26, 2018. The Carr fire vortex, though not seen by RaDFIRE, taught us a few new things about fire vortices.

The National Weather Service estimated wind speeds of 143 miles per hour in it, which makes it a true tornado of class EF-3. (California, a state not known for tornadoes, had seen only one of them reach class EF-3 in its entire prior history.) It was seen up close by plenty of people. Some shared phone-camera movies of it online. It uprooted trees without singeing them, proving that once a fire sets a twister in motion, it can spin off, independent of the fire.

You may have seen aerial photos of houses burned in other suburban catastrophic fires like Santa Rosa or Colorado Springs. Where there was a metal roof, it dropped crumpling onto the foundation like a mussed-up bedspread. Where the roof was flammable, it's gone, leaving the house foundation like a minimalist-sketched house plan with the major appliances scribbled in to indicate a kitchen or laundry. Washers, dryers, and fridges are twisted, blackened metal frameworks. There are plenty of houses like that in Redding, but not in the tornado's path. The tornado left clean, bare foundations. Refrigerators were blown blocks away before catching, perhaps, in the tangles of a collapsed high-tension power line tower. A few high-tension towers were themselves blown hundreds of yards, crumpled, and partly wrapped around trees.

For now, we have only very limited lessons to take home from the fire tornado. The clearest is that fires at some points in time cannot be fought; they can only be fled.

No one knows how to predict a big fire tornado. It's actually quite common for eddies or whirls of air to form, and the heat of a wildfire makes it all the more likely. A sharp bend in an advancing fire line can induce whirling. Two or more growing hot fires near each other will tend to induce vorticity in each other's convection columns, and clusters of spot fires do often ignite concurrently and grow rapidly in the vicinity of a fierce wildfire. If a whirling convection column (and that's a description that applies even to an ordinary dust devil) coincides with a strong, increasing heat source (an explosive fire), it will tend to stretch upward and narrow itself into a fire whirl. Jason Forthofer of the Fire Lab in Missoula, who wrote a review of what's known about vorticity in wildfires, thinks of fire tornadoes as simply the big-and-strong extreme end of the spectrum of fire whirls.

This tornado formed at the fire front, and the fire did not go much farther in that eastward direction. It may be that vortices don't tend to make fires spread farther. A vortex sucks air at the bottom, so the surface winds at its base suck fire in rather than pushing air forward. On the other hand, it can pick up huge burning objects and throw them around; it lifts larger firebrands, lifts them higher, and carries them farther than other kinds of wind can, starting new spot fires at a greater distance. The Carr fire spawned plenty of spot fires, both before and during the tornado.

When I look at aerial imagery of burned wooded neighborhoods a few blocks away from the Carr tornado path, I see a pattern familiar from other suburban wildfires: the house is gone, the trees within twenty feet of the house are blackened skeletons, but fifty or more feet from the houses the trees still have their twigs and leaves, and some may even have survived. In other words, the houses burned much hotter than the forest. They are often the fuel that jacks up the fire's intensity and launches firebrands that set other houses on fire. An analysis of how this tornado formed sees several factors joining forces: the preexisting winds and local topography, a pyro-cumulonimbus storm developing, and the addition of hotter-burning fuels as the fire hit groups of houses. All of that stood on a foundation of climate: that month was the hottest month on record in Redding, and the preceding winter had seen about half of the normal rainfall. (Climate models show Northern California continuing to get hotter, of course, but as for rainfall, they see not less of it than in the past but wider swings: wetter wet years and drier dry years.)

The night of July 6, 2017, brought scores of thunderstorms with scant rain to the Chilcotin Plateau, in British Columbia. In the morning, fire managers in Williams Lake woke up to reports of 160 new lightning fires. British Columbia has a massive fire suppression infrastructure, but 160 fires in one morning is beyond the scope its budgeters had in mind. The emergency call center line went down, swamped with calls, and the power began flickering as small power lines burned down in various corners of the grid.

Forest ecologist Greg Greene, nominally a research assistant at the University of B.C. Fraser Research Forest, fought fires for the rest of the summer. He had the skills, as he had spent the four summers of his college years fighting fires in New Mexico. Fire crews flew in from many of the western states and from Australia. Yet B.C.'s Summer of Fire blazed on.

It was still going strong on the afternoon of August 12 when a cold front brought unstable air—a recipe for forming pyrocumulonimbus clouds. Five separate pyrocumulonimbus clouds developed between two o'clock and five thirty that day, collectively lofting as much particulate matter into the stratosphere as a moderate-sized volcanic eruption.

I met Greene years ago at an alpine lake in the Valhallas, one of B.C.'s ridiculous number of spiky mountain ranges no one outside the province has heard of. Backcountry campgrounds in grizzly bear country are built with a common food area well separated from the tent sites; that way, it is hoped, food smells won't lure bears into tents. This has a happy side effect: it gets strangers to socialize over meals. Greene and I soon discovered our shared interest in forest science. A student at that time, now he has his master's degree and his dissertation data completed, and he lives in Williams Lake with his partner and their gurgling new baby girl. Kelly has a full year of maternity leave from her job as principal of the Horsefly School, forty miles to the east where the land begins to rise to the next mountain range. Greene is a fit thirty-eight-year-old with blue-gray eyes, close-cropped hair just starting to turn steely, a T-shirt that says "May the FOREST be with you!," and a cap that says STIHL—a chainsaw brand. Pairing those messages may have been inadvertent, but hardly ironic: Greene's studies and his own research firmly ground his belief in chainsaws as tools for forest health.

By the end of August 2017, twenty fires had coalesced into the Plateau fire complex, easily the biggest fire in British Columbia's written history, twice as big as the concurrent Hanceville/Riske Creek complex, which in turn exceeded the biggest forest fires in the lower forty-eight states during our lifetimes. By the end of the summer, around three million acres burned in British Columbia. In September one of B.C.'s fires blew across the Continental Divide into Alberta, where it completely toasted the forested portion of the western half of Waterton Lakes National Park. The total burnt acreage in the province roughly equaled that of 1910—the Big Burn—in Idaho and Montana. Thick tongues of smoke oozed south down the Fraser and Okanagan Valleys, generously sharing B.C.'s air-quality misery with Seattle and Portland for weeks on end.

Mysterious skinny, smoky columns snake a few hundred feet into the sky—a dozen or more of them in the course of the day—as we tour Hanceville/Riske Creek's devastation ten months after it burned. At first we think they rise from smoldering hot spots malingering from last summer and now rekindling themselves as the weather heats up. (Several such rekindlings were reported this month.) But these look too skinny for fire smoke, and as time passes they don't stay put, but wander a little across the landscape before fading out. They are dust devils, really tall ones—the incipient heat of summer playing games upon a ground surface covered with fine ash.

Mile after mile of burn goes by until we reach the zone we're looking for: all the burnt lodgepole pine snags arch over to the east-northeast in long graceful arcs, tops mopping the ground. (Just think how it might have affected Indigenous architecture if the forest had supplied not only lodge poles but lodge arches like these, back in the day.) Woodworkers know that if you bend wood by steaming it for a while before applying force, the wood retains the bend as it cools down. The moisture in these poles must have turned to steam as the wave of blast-furnace heat bore down on them and cooked them to the al dente noodle stage. (Well, at least in southern Italy they say al dente spaghetti is not limp; it arches.)

Here in the western part of the Chilcotin Plateau, white peaks and ice caps of the Coast Mountains loom tantalizingly in the distant west. We are in their rain shadow, where it's too dry for any trees to grow that don't go by the name "lodgepole pine," and even lodgepole grows slowly. Few trunks are even eight inches in diameter. Bigger ones may have been logged here in the past, but regrowth was taking its time. Probably lots of them were already beetle-killed before the fire. Because the timber volume here was meager, salvage loggers left it alone.

Mosquitoes swarm us as we get out of the car. While Kelly and baby

Madelynne stay in the car to avoid bug bites, Greene and I and Costner the dog go looking for any organic layer on top of the soil; but it's all gone, turned into a thin ashy residue already disappearing into the sand. Or going up in dust devils.

On August 11, 2018, a year to the day after the pyrocumulonimbus event, British Columbia again woke up to 140 new lightning fires in one morning. This broad weather front sparked an additional sixty-eight new fires in Washington and many more in Idaho and Montana, including one that shut down the highway through Glacier National Park. Week after week of hot dry weather across the region had dried all dead biomass to around the moisture level of kiln-dried lumber. By the end of the season, British Columbia saw 3.1 million acres burn in 2018, slightly exceeding 2017 and exceeding the annual average from the preceding ten years by a factor of eight. In terms of fire damage to communities and property, 2018 was milder than other years, but in smokiness, and smoke's effect on human health, 2018 was the worst year anyone can remember.

Increasingly, our forests fail to come back after fires. In the parts of the West where ponderosa pine is common, more than one-third of the area burned in this century has zero or negligible conifer seedlings coming up.

Studies in different parts of the West have come up with strikingly similar numbers. A study of fourteen fires spread across Northern California found that 41 percent of random plots had no seedlings within five years; it was worse—63 percent—on sites that had been ponderosa/Jeffrey pine–dominated; and on sites that had been mixed-conifer, most of the new growth was fir or cedar, so a sharp decline in pines was underway. A study of eight fires spread across Arizona was even more ominous: these were replanted with ponderosa pine in years that got almost normal precipitation. Of the eight sites, seven failed to regenerate naturally, and in four even the planted seedlings failed to survive.

The broadest recent study covers fifty-two fires in the northern Rocky Mountains, analyzing data from eleven previous publications. It found that within the dry forests (the ones with ponderosa pine), 47 percent had no seedlings. In all of these studies, centers of large high-severity patches were the least likely place to find seedlings.

It's easy to guess why this is happening.

For the most brutally simple, all-encompassing reason, just look at climate change. The ranges of tree species are moving north, and uphill. Ponderosa pines (along with piñon and juniper in the Southwest, and limber pines or Douglas-fir in a few places) are the trees of the driest habitats that can support trees in the West. That's their zone. Logically, when the climate on those present-day forest margins gets one notch hotter and drier, it pushes marginal habitats over the edge. They become habitats that cannot support trees. We have to expect the lowest margins of forest growth to shift northward and upslope, and the trees currently growing at those margins to die off. In a central Idaho study, new pine and Douglas-fir seedlings are reasonably common on burn sites with relatively mild summers, but are generally missing from burn sites with mean summer temperatures over sixty-three degrees Fahrenheit for Douglas-fir, or sixty-six degrees for ponderosa. Most sites in the study are projected to be too hot for any Douglas-fir seedlings by midcentury. Similarly at Yellowstone, conifers fail to regrow at lower elevations after fires, given temperatures as hot as we expect. Meanwhile, the highest elevations in both studies appear to be getting more favorable to seedlings of both species.

Fire can serve as the catalyst for climate-driven shifts in forests. There have been times of rapid change in forest composition before, and in the West they often coincided with centuries with a lot of fire. The biggest such shifts came when the globe warmed rapidly after the last ice age, and then again during a more gradual climate shift about six thousand years ago. (See chapter 7.)

Where forests are able to return, many of them will have fewer tree species than before. While no one can predict exactly what will be in all this "nonforest" that's going to replace many forests, it's safe to say that along with (or in) some novel communities we will get lots of plants that we see now in nearby arid nonforest habitats:

- Grasses
- Sagebrush grassland-steppe
- Chaparral in California and nearby parts of Oregon and Arizona
- Gambel oak and New Mexico locust in Arizona and New Mexico
- Mixed-species brush communities in the southern Rockies
- Weedy non-native grassland in California
- Weeds generally, including massive invasions of single species

Old trees growing today near the margins of forest growth survived multidecade droughts from time to time during their lifetimes. Seedlings, however, rarely survived those droughts. They tolerate far less heat and drought than grown trees do. Because the ground surface heats up in the sun, in turn heating layers of air near the ground, the hottest places in sunny terrain are about two inches above the ground. They often exceed 140 degrees Fahrenheit. That's no big deal to a thick tree trunk, which cools down each night and takes many hours to heat up again behind its insulating bark; its living phloem and xylem tissues virtually never reach 140 degrees. A pencil-thin seedling, in contrast, gets no such benefit from its own mass. Every day it heats up pretty close to the ambient air temperature. There's a limit to how much heat that little stem can take. If the hot zone two inches from the ground exceeds its tolerance limit, the seedling dies.

One reason ponderosa pine can be a tree of the hot and dry margins is that its seedlings can cool themselves by as much as thirty-five degrees Fahrenheit by maintaining a high rate of transpiration through the needles. That draws lots of cool water up from the soil, cooling the stem. It only works as long as the roots can draw moisture out of the ground, so a lot of the time it can't help. In moist soil, a ponderosa pine seed can grow a twenty-inch root within a few months, but without moisture a first-season root is much shorter.

When a marginal forest falls off the climatic edge, the existing trees commonly hang in there until a fire or something comes along to kill them. No post-fire seedlings come up to replace them, or some come up but they don't survive. The site has converted to grassland or brush. So the expected shifts northward and upslope are likely to be lurches, coincident with fires or other mortality outbreaks such as bark beetles and fungal disease.

Broad areas of quaking aspen trees have died in recent years near the dry margins of that species' range. This mortality has its own name, Sudden Aspen Decline, or SAD. Most likely it's simply range shrinkage due to drought; other proximal agents of death have been seen, but aren't well quantified or understood. Aspen is North America's most widely distributed tree, and in much of that enormous range it's doing fine, notwithstanding chronic struggles with browsers (of the hoofed kind). On some Colorado mountains it's taking over from conifers after beetle epidemics, and in Wisconsin it's growing faster than ever thanks to carbon dioxide enrichment of the air. After the 1988 fires at Yellowstone, it was able to replace conifers in some severe fire patches, especially at elevations higher than where it had grown before. Overall, aspen is not one of the more vulnerable tree species. It's climate-caused trends will vary by region; big increases are projected in central Alaska.

Consider the immediate effects of fires. The hottest fires can devastate the terrain, leaving no organic soil, nothing to retain moisture after a rain, no shade, few viable seeds, few living fungi in the soil. (Most trees—actually, most plants—are more or less dependent on mutualistic relationships with soil fungi.) "When trees die from bark beetles or drought stress," Craig Allen tells me, "what happens afterward is quite different from after you incinerate them and the biomass and the seed pools and the soil communities are gone, the whole nine yards."

Intense fires infuse soil surfaces with fire-formed waxes so that rain runs off them like off a duck's back. The word for this is "hygrophobic"—afraid of moisture. Some soils lose not only their organic matter, which combusts, but also their fine mineral particles. Locally, Oregon's Biscuit fire turned the soil surface into sharp-edged parking lot gravel. Apparently the convecting hot air vacuumed up all the fine particles near the surface and blew them out to sea. At least two inches of soil took flight, as shown by charred root crowns now elevated above the new gravel surface. (The depth of organic soil lost can be far greater, like ten inches across much of the King fire.)

Of course, high-severity fires have been happening forever, and forest almost always grew back. What changed? For one thing, there's considerable latitude in the term "high severity." All that we know about presettlement high-severity fires, based on tree-ring study, is that they killed most of the trees in the stand. We can't know whether they burned this hot, or did this much soil damage. Recent fires may be achieving greater heat.

Second, high-severity fires were common not in ponderosa pine but in wetter higher-elevation forests. Those trees are better adapted to recover after fire. Since the places they grow in are cooler and wetter, desiccation of new seedlings is less of an issue (though it may become one as climate warms). In contrast, high-severity fires were not common in the dry-zone frequent-fire forests—the ones with a lot of ponderosa pine, Jeffrey pine, western larch, or sequoia. These trees have an entirely different fire strategy: they foment frequent surface fires and survive those, thereby making hotter, taller fires unlikely and rare. Reestablishing after a firestorm is not part of the plan their genes hold for them.

Even in ponderosa pine, there was occasional high-severity fire in the old days, and forests usually came back. They must have, or there wouldn't have been so much ponderosa pine forest in 1880. In Arizona, New Mexico, and western Colorado, conifer forests predominate, but brush fields of Gambel oak can persist for hundreds of years, with or without fire.

A key change we're seeing now, beyond the total acreage of ponderosa that burns at high severity, is the size of the high-severity patches. Near the Las Conchas fire, photos from 1935 show brush fields whose outlines didn't budge over the next seventy-five years. They likely originated as high-severity fire patches. The biggest Las Conchas fire high-severity patch that is now becoming an oak-locust brush field is thirty times larger than any of those older patches.

For predicting forest recovery, the distance to the nearest surviving seed trees is the single most useful metric. If the dead tree patch is huge, even if the soil isn't badly damaged, the conifer seeds for a new forest won't get there for one or more centuries. (Dead lodgepole pines or knobcone pines can be terrific seed sources with their serotinous cones that outlast fires, but other species rarely match that trick.)

Another example of failure to regenerate after fire can be seen in the mountains east of San Diego, where California's third largest wildfire, the Cedar fire, burned forests of Jeffrey pine at the lowest, hottest margin for that species. In the fifteen years since the fire, Jeffrey pine has been replanted in three different years, and hardly any seedlings survived. Growing in its place, so far, is a community of shrubs, a lot of non-native weeds, and a few Coulter pines.

But how do we know pine forests won't grow back if we wait forty or eighty years? We don't. In many cases they probably will. Several forest ecologists are hard at work to answer exactly that question. Some tree-ring studies in especially dry corners of pine range, in the Southwest, conclude that episodic regeneration has been the norm for centuries. Seedlings germinated and survived most often after three years in a row came along that were wetter than the long-term average. Presumably those trifectas had to also coincide with a local mast year—a year when the pines produce a lot more cones than usual. These studies found that post-fire landscapes literally often waited three or four or five decades for that favorable combination to occur. Still, if it's a matter of waiting for a three-year spell cooler and wetter than recent norms, and if forty years from now the norm is another degree warmer than 2018—currently the hottest summer on record in the West—the odds look dicey.

Federal budgeting processes are not a good fit with a natural system of episodic reproduction that happens just during occasional groups of wet years. After the Las Conchas fire, funding was approved to plant 425,000 seedlings in selected areas. When the seedlings were ready, ecologists saw that it was a bad year to plant them, a year of extraordinary drought. They objected, but the replanting funds had to be spent in the year they were budgeted for, and the nursery had to move on and grow other seedlings, so the 425,000 seedlings were stuck in the ground. Ninety percent died. No further replanting on that scale has been funded there.

California is known for its chaparral ("shap-a-RAL"), a particular type of dense brush field made up of mixed species, most of which are broad-leaf evergreens. Very similar communities grow in other Mediterranean climates under other names—*macchia, maquis,* or *fynbos.* These communities quickly regenerate after fire. Many of the species can resprout from the root crown after the aboveground part of the plant burns. Some have seeds that germinate when stimulated by heat, even after lying for hundreds of years in the soil. In other words, forest can take over and shade those species out of existence locally for hundreds of years, and they can still spring to life after the forest burns.

Forest and chaparral have been competing more or less forever in the Sierra Nevada. Chaparral gained the advantage during the gold rush era, with its fires and clear-cuts. The fire exclusion era returned the advantage to the conifers. In some cases, especially on south-facing slopes, where forests burn today and chaparral takes over, you could consider it a return to more natural vegetation.

California's Eldorado National Forest is at the heart of the unending turf war between chaparral and forest. As its name implies, it was hit hard by gold-rush miners. In the past ten years it saw the King fire and two preceding large fires. Reforestation here is overseen by Dana Walsh, the district's energetic young silviculturist. Walsh resembles a ninety-five-pound forest sprite in a forest-themed baseball cap and a motorcycle-themed T-shirt. "A lot of the Eldorado comes back shrubs very quickly," she told me. "I have no doubt that we will have another fire. Our goal is to keep the shrub competition low enough to let the trees grow through the fire risk period quickly" to where they are tall enough to survive a low fire in respectable numbers.

In between the snags there's already a carpet of young chaparral shrubs, notably whitethorn ceanothus, which can turn a pleasant stroll bloody. Chaparral will be the fate of most federal land in portions of the fire that don't get cleared and replanted within a few years. (Not all of that is *conversion* to chaparral, as there were already expanses of chaparral before the fire.) Walsh selected the ten thousand federal acres best suited for reforestation, but how much of that plan gets carried out is subject to the vagaries of fate and funding. "In 2016 we ended up with a huge wet year. A lot of rain on a large fire area can do a lot of damage. We lost the entire road system to the northern portion of the fire. We couldn't do any salvage logging there that year, and now the wood no longer has any value to it, so we have to pay somebody to harvest that wood."

Finding ways to dispose of that no-value biomass is a huge problem. (I'll go into that in chapter 6.) Finding people to do the work is another. "In the national forest we did a lot of reforestation in the 1990s, following the eighties' clear-cuts. After that our reforestation shifted and really slowed and went to following major fires, and the agency lost a lot of that knowledge, a lot of that capacity. The number of contractors out there is really diminished. Even if I had endless amounts of money, the ability to get the work done would be limited by capacity."

Meanwhile, about half of the King fire area is private industry land (even though all of it shows up as Eldorado National Forest on your highway map). Private industry forest, after a fire, gets stripped completely clean and turned into a new plantation. Shrub competition is wiped out with aggressive herbicides and sometimes deep tilling. The Forest Service uses herbicides for shrub control as well, but aims for relatively incomplete and short-lived shrub reduction. "My goal," Walsh tells me, "is about thirty percent [maximum] shrub cover for ten years."

Some of the issues (especially the budgetary ones) might be alleviated in the near future by replanting with drones. Drones can't plant seedlings, and they can't even plant seeds, they can only drop seeds. That's a disadvantage they might be able overcome with sheer numbers of seeds dropped at lower cost. The company testing and refining drones for replanting claims that they can search and find the optimal spots for planting, and they can pick individual shrubs and spray them with herbicides first, avoiding spraying the entire project.

"For me in the Sierra Nevada, reforestation isn't just putting trees in the ground. Along the Oregon Coast it's not water-limited, you can just put trees in the ground and let them grow; but here, we've got to do shrub control, we've got to do fuels reduction."

The genus *Ceanothus* has fifty-five species, forty of them in California, and they fix nitrogen, fertilizing the soil and potentially helping conifers return. In central Oregon the post-fire *Ceanothus* species is an attractive aromatic one called snowbrush or tobacco brush. Foresters used to see it as an obstacle to reforestation, but Jerry Franklin tells me they now find that conifers can come up through snowbrush in the eastern Cascades. California chaparral is different; it's a serious, long-lasting obstacle to conifers, especially within large patches of high-severity fire. In some places white fir manages to grow up through it, being shade-tolerant, but the other conifers do not, and we'd rather see those other conifers, because they're more drought-tolerant. In recent decades, we're finding that where chaparral takes over, it is likely to burn severely again before conifers can regain dominance.

In other, still harsher parts of the West, conifers may actually *need* brush to come first, to provide shade to help conifer seedlings avoid desiccation. At Mesa Verde, for example, park ecologist Tova Spector told me she thinks a weedy, desertified severe burn will take fifty years to become a good brush field, and then another fifty to return to piñon-juniper woodland like it was before the fire—if only it can get through those one hundred years without a fire.

While reconversion from brush to forest does happen, the outlook overall is not rosy. Something that worked in 1850 may be impossible in 2050, given that the Southwest's climate has warmed only a degree or two so far, and we're headed for several degrees. Further warming is baked in by the quantity of greenhouse gases already released.

Forest edges getting too hot and dry to support forest doesn't happen only in the Southwest; moisture-limited forest edges lie at the foot of nearly every mountain range in the West. They constitute the lower timberline, the boundary where the first trees appear as you ascend from the land of sagebrush or of crops. The pattern of a fire marking the moment when the lower timberline lurches upward, when forest converts to shrub-steppe or grassland, has been seen in Idaho, Wyoming, Montana, Washington, and Northern California.

Washington's Chelan and Entiat Ranger Districts have seen well over half of their acreage burned since 1988, much of it twice. In 2018, the Cougar Creek fire reburned forty-one thousand acres of the 1994 Tyee Creek fire in steep, rugged mountain terrain. Before 1988, the Chelan and Entiat districts were forested, so richly forested that we know this rate of burning cannot have been the norm. The biggest alpine larch specimens grow here, hundreds of years old. Now, entire viewsheds are mixtures of brush and nearly barren, with few tree seedlings surviving. Time will tell whether forest will grow here again.

The 2002 Hayman fire was Colorado's largest and most severe fire to date. It blew up on its first day, burning fifty-nine thousand acres in twenty-four hours, and left a core treeless patch of fifty thousand acres. So far, conifer seedlings are sparse to nonexistent in the big treeless patch, and the return of ponderosa pine forest to that patch is predicted to be a matter of decades, centuries, or never.

In the Klamath Mountains of northwest California and southwest Oregon, a recent study posited a new wrinkle to the picture of high-severity fire patches too big for the conifers to reseed from the edges. The Klamaths hold the broadest U.S. expanses of the mixed evergreen forest type, i.e., where evergreen conifers mix with evergreen broadleaf trees. You can probably picture some example of broadleaf evergreen leaves, depending on where you live, perhaps rhododendron or magnolia. The Klamaths have a lot of those, tree-sized ones fifty to a hundred feet tall, called giant chinquapin, madrone, tan oak, live oak, California bay. They are as flammable as conifers overall, unlike deciduous broadleaf trees, which often function somewhat as firebreaks. These broadleaf evergreens have an excellent fire strategy: after a fire kills their aboveground parts, they resprout from their root crowns. That's faster and surer than growing from seeds. In the current climate they tend to take over toward the centers of high-severity patches.

Broadly, across the West you can expect an increase in fire to favor resprouters, most of which are shrubs. Gambel oak, in the Southwest, is a resprouter flexible enough to grow as either a tree or a thicket of shrubs. In the Klamaths, resprouting broadleaf evergreen shrubs abound, both underneath forest conifers and as chaparral communities.

Apparently conifer seedlings in the Klamaths can hold their own and eventually overtop the broadleaf trees, but *only* where their seeds are plentiful in the first two or three years. When fire next burns this new broadleaf forest, the conifer seeds will be scarcer than ever, so the broadleaf stand will perpetuate itself. The study, by Alan Tepley and colleagues, predicts extensive type conversion: "By the end of the century more than half of the area currently capable of supporting montane conifer forest could become subject to minimal conifer regeneration in even moderate-sized high-severity patches."

The Klamaths get plenty of rainfall, at least in the winter half of the year; unlike most predicted conversions from conifer forest, this one is nowhere near the arid margin of where these conifer species can grow. We can perhaps take a degree of comfort in the fact that this conversion isn't a disappearance of forest, just a change in the kind of trees, with a somewhat lower level of biomass. Also, what the conifers need in order to be able to carry on is simply to germinate in the first couple of years post-fire, so the conversion could possibly be averted by planting seedlings. On the other hand, a modeling study found that a climate expected to convert one-third of the Klamath region's conifer forest to broadleaf already existed by 1990.

In British Columbia, it may seem to stretch credulity to say that it's getting too hot for forests. Yet even there, reforestation on drier sites is starting to look dicey. Douglas-fir seedlings may not survive unless they can get some shade from another species establishing first, such as lodgepole or ponderosa pine. The issues are more complex, but they still revolve around what a tree seedling may experience as drought stress. As Craig Allen explained to me, trees suffering from drought are not limited to the world's dry places; it's all relative to whatever climate the particular tree grows in. "A juniper in Arizona would laugh at what feels like a drought to a tree in the Amazon. A tree in the Amazon is tuned to compete in that wet environment, so its tolerance to drought is much less."

Fire in the West is on the rise. The increase since 1980 has been hard to miss. The hotter climate is directly at fault, and so, in proportions that vary by site, are forest density and homogeneity. While the number of annual fire starts has not increased much, the number of large fires and the average size of fires, and therefore the total acreage burned, have shot up decade after decade. The upward curve in the dollar cost of fires—including both fire-fighting and rebuilding—looks exponential.

Taking a more ecological view, the proportion of burned area that was deforested (i.e., high-severity fire) has increased in California, and perhaps on the Cascades east slope. Elsewhere in the West, lodgepole and spruce-fir forests are a bigger part of fire trends, and they have always burned severely, so the high-severity proportion hasn't risen much. What's ominous ecologically is the increasing area in large deforested *patches*, and the proportion of those that do not appear to be regenerating as forest. With warming, all three trends in spruce-fir forest—greater area burned, patch size, and failure to regenerate forest—are expected to sharpen.

However—and this is a huge *however*—the increase in the annual acreage burned is in comparison with the mid-twentieth century, the fire exclusion era, when acreage burned was very low: anomalously low, unnaturally low, nonrepeatably low.

For a meaningful comparison, we want to look at the acreage burned before 1850. We don't want to bother with a certain graph that's been all over the internet showing very high rates of fire in the 1930s through 1950s. That graph is based on crude data from the National Interagency Fire Center, but the Center disavows it as being wildly exaggerated. It includes huge areas of agricultural field burning in the Midwest and South, it double-counts many fires, it does not even have any reliable data to start from, and it is rebutted in some areas by more reliable data.

In contrast, there's an excellent three-thousand-year fire history of the West assembled by Jennifer Marlon with many coauthors who have been doing both charcoal-sediment and fire scar studies throughout the West. They found that fire was far more constant and extensive over most of the three thousand years than it was in the twentieth century. Further, "burning generally increased when temperatures and drought area increased, and decreased when temperatures and drought declined." Burning reached a low point—

not quite as low as 1970—in around 1600, with the onset of cold centuries known as the Little Ice Age. As climate warmed it rose to a moderate peak around 1850, when miners and settlers arrived, and after that it tanked, despite continued warming. In other words, it disconnected from climate and became human-controlled instead. Granted, Native Americans contributed at least a little bit to the generally high levels of burning before 1600, and their post-1492 decimation may have been a factor in the fire dropping to a low at 1600, but their role can't really be measured.

This may be hard to take, but the bottom line is that there has not been way too much fire recently. If anything, there's still too little fire, by long-term standards. As Marlon puts it, we have not yet caught up with the fire deficit.

In the ponderosa pine and mixed-conifer forests, it's a deficit specifically of low-severity fire. The area burning at high severity has regained its presettlement normal, but the area burning at low severity falls far short.

In any case, fires in increasing number and size are not optional. Our options now concern what kind of fire we get, not fire versus no fire.*

---

* I'm leaving the wet forests of the Northwest Coast out of this discussion, as I leave them out of the book overall. They look different, so far: no terrible insect scourges, no big fires since the 1930s. Drought mortality was not obvious here, at least until around 2018 when British Columbia ecologists pinned a wave of redcedar and salal mortality on drought. (A 2019 study describes microclimate mechanisms that may leverage sweeping ecological change during near-future warming.) They don't burn often or easily, and their natural fire regime ranges from moderate to high severity, but usually patchy. Lightning unaccompanied by rain is uncommon, so a high proportion of fires are human-ignited, and we can barely even speculate as to how many fires would happen "in nature," without ignition by humans. Their huge severe fires of the early 1900s were incubated in clear-cuts laden with slash, and may not be representative of natural patterns. They do not benefit from either prescribed fire or fuel reduction treatments, except perhaps toward the south end where they grade into frequent-fire-type forests. (The most beneficial treatment might be to bring greater heterogeneity to plantations that are typically single-aged Douglas-fir monocultures.) Fire suppression has not done major harm, and indeed, Jerry Franklin and colleagues support fire suppression as part of a program including some managed fire. These are among the very best forests in the world at sequestering carbon, and we maximize their carbon storage by just letting them grow, and not letting them burn.

The fire exclusion era looked like it was working, for a while, but then it brought about its own demise. It ended not because we stopped trying to put fires out but because hotter summers together with the accumulation of fuel make it inevitable that some fires will defeat our efforts, and those fires tend to get huge. Even fires that could be controlled individually will sometimes go uncontrolled because there are too many fires burning at once and not enough firefighting resources to go around.

An argument is sometimes made for maximizing forests' carbon storage by logging them at strictly sustainable rates, at very long rotations that allow them to mature, and then ensuring that some of their wood ends up in very long-lasting buildings or, better yet, recycled through more than one generation of buildings. The manufacture of steel and (especially) concrete emits huge amounts of carbon, practically begging us to find lower-carbon replacements for them. Can wood meet that test? Conceivably, but it's a stretch.

If we aren't seeing a too-much-fire problem, then what is our problem with fire?

We don't like losing homes and lives. We don't like spending ever greater sums of taxpayer money saving homes and lives, and in the process losing firefighters' lives—while not doing the forest much good.

We don't like smoke. The health hazards of breathing wildfire smoke are serious. Unfortunately, along with not having the option of no fire, we don't have the option of no smoke. Fortunately, the kind of fires we want to choose because they are better for the forest are also less smoky and less health-damaging.

What about the severity of the fires—the proportion of trees they kill? It isn't a problem everywhere. Some parts of the West just naturally grow types of forests that don't have high rates of survival when they catch fire. In other words, moderately high to very high in severity is the only way they can burn.

And they will burn sooner or later. Many people think of the 1988 Yellow-stone fires as good examples of too severe or too big. They aren't. Enormous fires had affected those areas in the past; the 1988 fires burned in the way lodgepole pine forests typically burn; they were not strongly affected by pre-vious fire exclusion; and their ecological effects were mostly good.

But we don't like so much high-severity fire in our frequent-fire-adapted forests with their irreplaceable big, old trees. We are blessed to have tree species that evolved exquisite adaptations to frequent, low, surface fires, en-abling them to grow into exceptional groves with big, old trees. These forest types are common across the West, and they are favorites with people. We are losing a lot of them because we denied them their frequent low fires for so long that fires now endanger them (and bark beetles do as well). Those forests—ponderosa pine, western larch, Jeffrey pine, giant sequoia—are the ones where high-severity fire is uncommon naturally, and undesirable.

We don't like the massive pulses of erosion that typically follow large high-severity fires, as the soil-gripping tree roots die and rot away. Massive erosion is an immediate threat to water treatment plants and a long-term threat to reservoirs and dams: the reservoirs fill up with sediment. Even-tually they will completely fill up with sediment and their dams will no longer serve any of the purposes we built them for—water supply, flood control, recreation, and electric power generation. The two clearest climate predictions regarding precipitation in the West are that we will see more floods and more summer water shortages. Both of those trends increase our reliance on reservoirs. The more big high-severity fires we get, the more we'll have to consider dredging sediment back out of reservoirs for them to function again.

In the frequent-fire-adapted forests, the uncharacteristic large high-severity fires threaten to reduce the total area of forest, the total number of trees, and their ecosystem services. Forests support wildlife, including economi-cally desired species like salmon. They help a lot with hydrology, preventing floods and keeping our water supply clean.

And then there's that huge, overarching service trees perform: they pull carbon dioxide out of the air and store it via photosynthesis. Molecules of water (H_2O, two atoms of H and one of O) from the soil and molecules of carbon dioxide (C + O + O) from the air are taken apart in the plant's chloroplasts and reassembled into various carbohydrates (Cs + Hs + Os) including sugars, starches, cellulose, and lignin. Excess oxygen is released into the air.

All green plants do this. Crop plants are especially efficient at sucking up carbon dioxide, but they don't hold on to it for long enough. Trees are more effective because they are bigger, and because a lot of their carbohydrate production goes into cellulose and lignin. Those compounds give wood its strength, and are far longer-lasting than sugars and starches. Indeed, trees are an enormous component of the global carbon cycle: the carbon pulled out and stored by the world's forests every year is equal to (and compensates for) more than half of the carbon we emit from fossil fuels. In the very long term, this is only temporary carbon sequestration; the carbon is re-emitted when the carbohydrate molecules rot, but a lot of forest carbon is pumped through roots into the soil, where it stays for a good long time before rotting. Charcoal from fires sticks around longer, a medium-term carbon sink.

When forests are on the increase globally, they substantially help with climate change. Miraculously, that seems to be happening over recent decades: climate change would be noticeably worse than it is if it weren't for what forests are doing for us. Scientists are puzzled as to exactly where this verdant carbon sink is located. Is it the reforestation of former farmlands, like in eastern North America? The afforestation of terrain that was formerly too cold for trees, like in Tibet and Alaska? Or the effect of atmospheric carbon dioxide fertilizing tropical forests (and plants in general) and making them grow faster, even while so much tropical forest is cut and burned? All of the above, but we aren't sure of their relative contributions. Satellite imagery's big picture confirms that the world is getting leafier; the green terrestrial biosphere is growing.

Western pine forests haven't been doing as well as some forests globally: bark beetles, forest fires, and harvesting have turned regions of western forests into carbon sources at times, exacerbating climate change. British Columbia's forests were calculated to be a net carbon source (putting them on the bad side of the balance sheet) during the pine beetle epidemic, and again in 2017 and 2018, the worst fire years. California has also likely become a carbon source since 2000.

For carbon's sake (among other reasons) we can say "the more trees the better." Unfortunately, in the warming climate, the arid West can't sustain as many trees as it did before. Many of its forests are too dense for their own good, and trying to keep them dense for carbon's sake would fail. Over much of this area some drop in the number of trees is both inevitable and desirable.

In consequence, we should appreciate most fires that thin the forest— low- to moderate-severity fires. Fires that convert a site from forest to grassland or brushland are tough: we lose the ecosystem services of a forest, including carbon sequestration. We face the reality, though, that some of those deforestations are already happening; in some places climate has already changed to one that doesn't support forest, and nothing can be done about it.

Sugar Pine

STRIEBY

3 OUTBREAK

Most of the lodgepole pine forests of Colorado and southern Wyoming are now standing dead . . . It is astonishing to me that so many people outside the west have no idea this has happened.

—ANNIE PROULX, *Bird Cloud*, 2011

The troubles afflicting western trees count fires and insects as their agent of death in roughly equal measure. Of the insect part, it seems 90 percent of the deaths are inflicted by bark beetles. Like fire, bark beetles belong here. They are natural. What alarms us is the upward trend. The beetle epidemic of 1999–2009 in the Rocky Mountains appears to have been about ten times bigger than any previous ones since 1880, which is about as far back as we can quantify them. To the best of our knowledge, the trend stems from warming climate and from fire exclusion.

People still benighted after that beetle epidemic may have woken up in 2016 when the headline became "Sixty Million Dead Trees in California." By the end of 2018 the Golden State's dead tree count was up to 147 million. Beetle-killed lodgepole pines in the Rockies doubtless number in the billions, but California's dead sugar pines and ponderosa pines may total more wood per acre, as they were big, stately trees, many of them hundreds of years old and two hundred or even three hundred feet tall. Trees well worth counting one by one . . . if you can count to 147 million.

For years I'd been seeing green slopes mottled with telltale red-brown in Wyoming, in Idaho, in California, and on and on. Those landscapes of death were nothing compared to British Columbia. Look at any of the maps of beetle-inflicted mortality on this continent in this century, and the terrifying big brown blotch is interior British Columbia. By far the biggest patch of pine mortality, an area bigger than Wisconsin, happened there, first appearing in 1990, peaking around 2006, and killing more than half of the pine timber volume in the entire province. Flights of beetles during the British Columbia outbreak formed swarming dark clouds.

Mountain pine beetles had attacked lodgepole pines in B.C. many times before. This time they spread farther north, where major outbreaks of mountain pine beetle had never been seen because up till now it was too cold. They killed even higher percentages of the trees there, because these trees were "naive": they were not genetically programmed to devote much of their resources to beetle defense. Now, the epidemic blazed northward to a latitude with a terrible significance, the fifty-fourth parallel, the southernmost latitude where the conifer forests of western and eastern North America meet. (South of there, hundreds of miles of treeless prairie separate eastern and western conifers.) Specifically, lodgepole pine meets its close cousin jack pine near the fifty-fourth parallel in Alberta. On a few windy days, millions of pine beetles literally blew over the Continental Divide. The mountain pine beetle met jack pine, tasted it, and it was good. The mountain pine beetle scourge may now be able to spread eastward, to Michigan's great forests of red pine and white pine, or even to the Atlantic.* Whether that will actually happen is the subject of much debate among scientists. It's an interesting debate but largely speculative, and since it is outside the scope of this book, I'll just say "we shall see."

Most of the time, however, bark beetles are barely noticed. An adult pine beetle resembles a grain of rice, dark reddish to black, with tiny wings that it can open up to fly with. (See the sidebar for a differentiation of "pine beetles" and "bark beetles.") As soon as it can decide on a good tree to make a home in, a female adult bores into the bark and begins chewing a vertical "egg gallery" in the tree's phloem, the sap-rich new-growth inner bark layer. A male spots the entrance she drilled, enters after her, and they mate in the gallery. She lays eggs one by one along the gallery's flanks; larvae hatch and begin eating their way into the phloem off at angles to the gallery. They chew in patterns that you can use to identify species of bark beetles once the bark is stripped off. In mountain pine beetles they are like short rungs; in some species the rungs start out close together and then fan out gradually, in arcs; in western pine beetles they curl almost as randomly as spaghetti in a bowl. The larvae overwinter under the bark, resume tunneling in the spring, pupate, and wait for hot weather before metamorphosing into adults and boring their way out into the daylight to take flight. Their many exits through the bark resemble buckshot holes.

---

* This climate-enabled northerly end-run route around the treeless plains was previously exploited, in the opposite direction, by another species connected to western forest issues. For decades, protecting the spotted owl, *Strix occidentalis*, has been a focus and a tool of forest activists. Significant protections were won, but in the end they may go for naught because that owl's closest relative, the barred owl of eastern North America, invaded the West via the northern Alberta forest route, starting in the 1960s. The barred owl is bigger, bolder, more aggressive, and less picky in terms of habitat and diet. It tends to displace spotted owls from the forests it moves into. To cap it off, the two are somewhat interfertile, so even if the northern spotted owl (subspecies) doesn't go extinct, it may end up hybridized out of existence.

TREE-KILLING BARK BEETLES

"Bark beetle" is a descriptive term for many beetles whose larvae live and eat under tree bark. Several of the species attack pines and are thus pine beetles. We have several thousand species of bark beetles, nearly all of which are pretty harmless, just hastening the decomposition of dead trees, or sometimes hastening the death and decomposition of terminally weakened trees. Around a dozen species are involved in the occasional population outbreaks that kill trees in large numbers. The genus name *Dendroctonus* means "tree murder," and beetles within that genus do commit the bulk of the tree murder by beetles that we're concerned with in the West. Here's a chart of some notable species:

COMMON NAME	SCIENTIFIC NAME	HOST TREES
Mountain pine beetle	*Dendroctonus ponderosae*	Lodgepole, whitebark, limber, sugar, jack, and western white pine; also ponderosa pine mainly inland
Western pine beetle	*Dendroctonus brevicomis*	Ponderosa and Coulter pine, mainly in the coastal states and B.C.
Jeffrey pine beetle	*Dendroctonus jeffreyi*	Jeffrey pine
Southern pine beetle	*Dendroctonus frontalis*	Ponderosa pine (AZ-NM); major pest of southeastern U.S. pines
Mexican pine beetle	*Dendroctonus mexicanus*	(Only recently came north across border)
Roundheaded pine beetle	*Dendroctonus adjunctus*	Ponderosa and other pines of AZ and NM

COMMON NAME	SCIENTIFIC NAME	HOST TREES
Spruce beetle	*Dendroctonus rufipennis*	White, Engelmann, and blue spruce
Douglas-fir beetle	*Dendroctonus pseudotsugae*	Douglas-fir
Red turpentine beetle	*Dendroctonus valens*	Fire-scorched ponderosa and other pines
Piñon ips	*Ips confusus*	Piñon pines
California five-spined ips	*Ips paraconfusus*	Ponderosa, Coulter, several coastal pines
Arizona five-spined ips	*Ips lecontei*	Ponderosa pine
Pine engraver beetle	*Ips pini*	Ponderosa, Jeffrey, sugar, lodgepole pine
Fir engraver beetle	*Scolytus ventralis*	True firs, especially red, white, and grand
Western balsam bark beetle	*Dryocoetes confusus*	Subalpine fir
Cedar bark beetles	*Phloeosinus* species	Incense cedar, junipers, Arizona cypress (It's unclear how lethal they are.)
"Twig beetles"	*Pityogenes plagiatus, Pityophthorus boycei*	Pines (This emerging threat remains poorly known.)

Wherever the larvae tunnel, they sever the tree's circulation. A tree that has beetle families thriving on all of its sides is girdled, and will die.

That life story makes it sound easy for beetles to murder a conifer. It is not. Trees defend themselves. Pitch flows to the initial wound where the female enters, and can quickly immobilize and embalm her. Or perhaps she gets in but pitch soon plugs the hole and engulfs the male that tries to enter after her. Even if they both survive, pitch flows in to infuse the wood in that entire area with toxic terpenes that can kill the larvae when they begin eating. (Pitch is not sap. Sap is a plant's water-based circulatory fluid carrying sugars and nutrients around throughout the plant. Usually it's mostly water, but in some conditions it's thin syrup, and you can boil it down to make syrup. Pitch, or resin, on the other hand, is soluble in turpentine or other spirits, not in water; it's a defensive secretion found in a different set of vessels called pitch ducts—and not in all plants, just mainly in certain conifers.) Within days, terpene levels multiply severalfold in the attacker's vicinity. The tree may also kill off a number of its own cells to create an inedible dead area.

Thanks to these potent defenses, the normal state of affairs is that most beetles die without multiplying. The population is sustained at a minimal ("endemic") level by a few beetles that select trees that barely resist them, so those few beetles reproduce successfully. Each female adult faces the enormous challenge of picking a tree that is too unhealthy to defend itself, and yet big and healthy enough to provide phloem food to a large brood of larvae. Typically these endemic-level victims are good-sized trees injured by a fire or lightning, or weakened by dwarf mistletoe, rot fungi, or various other kinds of bark beetles that are not life-threatening.

The beetles have chemical and biological weapons of their own. They inoculate the wood with fungi that they bring with them in a little pocket with just that function. The fungi eat deeply into the tree, producing fungal tissue in the tree's outer parts, where it becomes food for beetle larvae. Pine lumber from beetled trees often shows blotchy blue stains, a hallmark of the main fungal partners of the mountain pine beetle.

Adult beetles use smell to detect their victims. They also detect chemical signals emitted by beetles. If one female can get lots of others to pick the same tree, the tree may not be able to produce pitch fast enough to kill all of them, and many will succeed in mating and breeding; so after picking a promising tree, she emits aggregation pheromones that attract conspecifics. Within days the attackers can detect when they have won the battle and the tree is doomed. At that point they emit *anti*aggregation pheromones to repel others from joining, because before long there could be too many broods for the limited supply of moist phloem to feed all the larvae. Males, meanwhile, emit whiffs of disaggregation pheromones around their gallery entrances just to keep other males away from their mate.

Much of the time, even pheromone-assisted mass attack still falls short of producing a large population of beetles or killing a lot of trees. But at intervals of several decades, a region may develop a large population of mature trees of a single target species, and then if drought weakens them, the result can be a beetle population explosion, an outbreak, an eruption, an epidemic. An outbreak is kind of like that same group-attack, aggregation strategy cubed.

Outbreaks are especially pronounced in one species, the mountain pine beetle. Once its population explodes, gathering enough allies to overwhelm a tree's defenses becomes easy: even the healthiest trees are easily overcome, and since they also hold the biggest food rewards, they're the ones the beetles go for. As their reproductive success skyrockets, beetles soon spread to other healthy forests nearby. They no longer seek stressed or weakened trees, but actually prefer healthy well-defended trees, which are a richer food source; the mass attack is powerful enough to overcome any tree's defenses. So a mountain pine beetle outbreak is a self-reinforcing momentum thing, like a market bubble. Once it is going strong, there is absolutely nothing humans or trees can do to slow it down. Two known kinds of events can stop it: depleting the supply of good host trees, or a sharp turn toward colder weather. But the good cold snaps are getting few and far between.

Beetles have a chemical defense against cold snaps. They prepare for winter by synthesizing antifreeze proteins and alcohols with a low freezing point. In the dead of winter, protected under a good layer of bark, a mountain pine beetle larva can survive cold down to almost forty below. (Conveniently, that's the one temperature that's the same in Fahrenheit and Celsius.) Unfortunately for pine trees, it's a temperature that almost never happens. It takes the larvae weeks to get quite that pickled, though, and then they need a long gradual detox in spring, so a relatively modest cold snap in October or in April can kill them all just as effectively as forty below in January. Until the last thirty years or so, spring or fall cold snaps cold enough to kill most of the larvae did come along every few years in interior B.C. and in the U.S. Rockies at the elevations where most lodgepole pines live. Pine beetle populations would get reset to extremely low levels and pushed back to lower elevations, from where they would have to work their way back up.

Both for that reason and for one other, connecting the unprecedented size and location of the recent beetle epidemics to the warming climate is something we can do with confidence. The second reason relates to temperatures during the warmer half of the year. When they are warmer during their main growing season—or warm enough to mature, fly, and mate earlier in the spring—beetles complete their life cycle faster. High-elevation mountain pine beetles in the Rockies, or spruce beetles in Alaska, used to take two or three years to reach adulthood, but now they often do it in one. (Some other species, including the western pine beetle, complete two or more generations in one year.) Given that each female bears several dozen eggs, and that the population explosion process is one of sheer growth momentum, you can imagine what a difference it makes when a female has a potential for, say, 144 offspring in two years rather than twelve.

In Canada since 2000, a climate-related positive feedback came into play: as the beetles pushed northward into regions that had formerly been too cold for them to mount an epidemic, they fed on pines that were poorly defended because their genes had not experienced natural selection during past beetle epidemics. This seemed to make the epidemic all the more explosive in those regions.

The other likely trigger of this vast epidemic was a vast continuity of similar lodgepole pine forests. Fires had been suppressed for decades, and while fires don't have the same gentle effect in lodgepole that they have in ponderosa pine, they do at least create patchiness; the lack of fires led to a shortage of patchiness across the entire region this epidemic hit. Mountain pine beetles are short-distance fliers and poor dispersers. For them to mount a big outbreak it helps a lot to have susceptible host trees, of similar age and condition, spread continuously across the landscape mile after mile. And that's exactly what they encountered. Some people say that more logging would have created the right patchiness; it sounds plausible in theory, but unlikely. There was no shortage of logging in central British Columbia.

Bark beetles have predators of many kinds—mainly insects—and while predators do dampen population levels in a general way, they don't seem to have much clout with a beetle epidemic in full swing. Woodpeckers, the main larger predators, rarely multiply fast enough or move to an outbreak in numbers great enough to slow it or end it. Their territorial behavior also keeps them from gathering in large numbers. Sometimes they help, though. "Woodpeckers are doing a number on spruce beetle in southern Colorado," according to Barbara Bentz, a Utah-based Forest Service entomologist specializing in bark beetles. "Most likely the American three-toed woodpecker," which specializes in spruce beetles but also feeds on mountain pine beetle and various nonlethal beetles.

Forest managers have attempted to suppress incipient outbreaks with "direct control," meaning rushing in to log and debark all infested trees as quickly as they can be identified. Those efforts often fail. It's next to impossible to stay ahead of the beetles in an outbreak. To be effective, you have to log and debark trees soon after they are attacked, and before the adult beetles emerge. That requires a preternatural level of skill at detecting attacked trees.

Early in the twentieth century, foresters were desperate for a solution to bark beetles. The tried electrocution on trees that were conveniently close to high-tension power lines. They tried radio waves. They tried a military incendiary weapon called Goop. They tried a patented sticky goop called Tanglefoot. Of course they tried any number of poisons, some of which tended to kill trees. (Even then, advocates claimed it could be worth doing just to prop up the salvage value of the dead lumber, by keeping blue-stain fungi out of the wood.) And there was much preemptive logging of uninfected trees that were judged to be in the high-risk pool. The ability to pick at-risk trees was even less than the ability to pick attacked trees, so this preemptive thinning was as good as random.

In truth, the one halfway effective tool we have for fighting bark beetles is those pheromones they use. Scientists have synthesized both aggregation and antiaggregation pheromones for mountain pine beetle and for several of the other species. You can staple a little packet of verbenone on an especially valued tree and it will be somewhat protected for a year, and next year you can staple another. This method isn't going to be practical for a whole mountain range of lodgepole pines, of course. But whitebark pines are few enough and valued enough that their human fans have organized to hang thousands of verbenone packets in some stands. It's a fine day's volunteer work in beautiful subalpine habitat. Sadly, verbenone packets don't smell quite like their near-twin and namesake, the aromatic essence of Spanish verbena flowers, but at least they don't smell as bad to us as they do to pine beetles.

A pine beetle synthesizes verbenone by oxidizing one of the most abundant pine resin terpenes. Before oxidizing, this terpene ironically is the main smell that attracts beetles to pines. It's a puzzling paradox: beetles are attracted to the terpenes, which are supposed to defend against beetles; and beetles are especially attracted to individual trees that are less well defended. We don't know exactly how that works.

Aerial spraying of pheromones has been successful in suppressing a different bark beetle, the Douglas-fir beetle. That beetle is a more promising target because it doesn't fly as far or produce such uniform broad outbreaks, showing up instead as patchy mortality across the landscape. It has grown into a serious threat within the past decade—again because of stressed, dense forests in a warming climate. The disaggregation pheromone for Douglas-fir-beetle and spruce beetle is methylcyclohexenone, and has proven itself in packet form and also in some aerial spraying trials as far back as the 1970s.

I learned about that in interior British Columbia. I came to Williams

Lake to see beetle-killed lodgepole pine stretching from horizon to horizon. But I came too late: the lodgepole mortality had peaked twelve years earlier. Some of it hides among Douglas-firs whose growth sped up after the competing lodgepoles died. Purer stands of dead pines elsewhere were aggressively attacked by either loggers or fires. So instead of beetle mortality, it's the Hanceville fire that stretches from horizon to horizon while we drive across it for hours.

Greg Greene had just been out in the woods discussing an experiment to use "the repellent pheromone in combination with the attractant pheromone in a push-and-pull method at a landscape scale—using aerial spraying rather than stapling packets to trees. They would push beetles away from areas where there's veteran Douglas-fir trees and suck 'em into areas [containing lethal bait traps among] really young Doug fir, less than twenty-five centimeters diameter. The beetles don't really go for those, so your risk of mortality is very small." Conceptually, it reminds me of the sacrificial anodes drawing corrosion away from the steel of a ship's hull.

Lodgepole pine has a common name celebrating its straightness and a scientific name, *Pinus contorta*, celebrating its crookedness. It warrants both descriptions. Subspecies *contorta* grows along the Pacific shoreline, exposed to constant wind and frequent salt spray, producing a bonsai-like form, twisted and wind-flagged. Subspecies *latifolia* typically grows pretty straight, slender, and fast, making it an architectural wood of choice for teepees and log cabins alike. It ranges from the Yukon down the Rockies to Colorado and west to Washington—almost as great a range as ponderosa pine or Douglas-fir and overlapping them, but shifted northward. Subspecies *murrayana*, found mainly in Oregon and California mountains, becomes a more robust tree, especially in the Sierra Nevada. Curiously, though molecular study confirms the genetic separation of three subspecies, genes alone do not account for *P. contorta contorta*'s contortions; when planted inland, away from the salt and wind, it grows pretty straight.

The species is especially chameleonlike in its choice of habitats. In Oregon, aside from the seashore, it is best known for holding sway over pumice basins—concavities that Mount Mazama filled with pumice gravel during its cataclysmic eruptions 7,700 years ago. Other trees don't like the pumice, which doesn't hold on to water. Other trees having to struggle is a common denominator of lodgepole habitats: cold air drainage pockets, lava flows, mudflows, serpentine soils. In each of those places, you may find a lot of lodgepoles in a small area.

The wide range of *P. contorta latifolia* is yet another story. There—in British Columbia and the Rockies—lodgepole heavily dominates extensive forests, most of them at higher elevations where subalpine fir and Engelmann spruce are the main competitors. Lodgepole dominates heavily at Yellowstone, partly because there's no ponderosa pine there; within the Yellowstone caldera it especially monopolizes the rhyolite-based soils, which are chemically similar to its pumice soils in Oregon. Lodgepole was by far the most abundant tree all across the Northwest and northern Rocky Mountains for four thousand years of the early Holocene, a time that was drier than today and hotter in the summers, but cold in the winters. (Drier overall and colder in winter, hotter in summer could also roughly describe the difference today between Yellowstone and, say, Puget Sound lowlands.)

One key to its dominance is its remarkable and highly effective fire strategy, called serotiny. Lodgepoles are bad at surviving fires as trees, but they survive them phenomenally well as seeds. If even as little as 10 percent of the forest that burned was lodgepole pine, probably 80 to 100 percent of the seedlings that come up in the first three post-fire years will be lodgepole pine. They may be way too numerous for their own good, in fact, forming a dog-hair stand of small trees that will be stunted by overcrowding for years. In optimal conditions, lodgepoles can grow quite fast, but they seem to sabotage their own conditions.

Here's how the fire strategy works. Serotinous cones on lodgepole pine trees are sealed shut by a resin with a melting point of 113 degrees Fahrenheit. The seeds inside, viable for decades, are protected through a fire by the closed cone. The fire kills the pines but melts their cone-sealing resin; the cone scales open over several days or weeks, shedding seeds upon a wide-open seedbed.

Lodgepole serotiny varies within subspecies *latifolia*: many trees are not serotinous, and others have some serotinous cones and some nonserotinous, meaning the cones open while on the tree to release their seeds, as in most other pine species. (Occasionally, summer heat opens even serotinous cones prematurely.) Subspecies *murrayana* has an entirely different fire strategy and is not serotinous. It has thicker bark and survives many fires. Stands on pumice in central Oregon require a mixed-severity fire regime to sustain them.

The Sierra Nevada foothills do have a champion of serotiny, the knobcone pine, whose cones may stay closed twenty years or more. More knobcone pines live nearer the coast, as do their serotinous close relatives, bishop and Monterey pines.

An interesting consequence of the serotinous strategy is that lodgepole seedlings are extremely abundant after fires hit lodgepole forests in the Rockies, but less reliably common when similar forests are massacred by pine beetles. A big fire reinforces lodgepole dominance, but beetle epidemics may locally shift a forest toward other species such as subalpine fir, Engelmann spruce, Douglas-fir, and aspen. Though the conifers on that list each have their own species of bark beetle, the different kinds of beetle epidemics usually strike in different years.

In the driest part of British Columbia's interior plateau—its western side—lodgepole pines dominate, and suffered mountain pine beetle outbreaks in the 1930s, the 1970s, and the 2000s. Northeastward from there, greater numbers of spruces and firs intermix with the pines, and naturally increase in proportion during a mountain pine beetle epidemic. They should be left in place to increase the forest's diversity and its resistance to any one species of beetle. Typically, logging companies choose to clear-cut when they salvage log damaged pines, leading to new single-aged lodgepole forests and setting the stage for the next outbreak.

After the 1988 fires in Yellowstone, lodgepole pine came back very strong overall, though there were local shifts from lodgepole to meadow, to brush, or to aspen. A concern has been raised, though, that in the warming climate, fires could short-circuit lodgepole's fire strategy if they come at shorter intervals, as some models predict. If a cohort of lodgepole saplings were to reburn before the age of ten, it would not have any cones on it yet, and the cycle of lodgepole reproduction would be broken. Big portions of the Yellowstone ecosystem might convert from pine forest to sagebrush grassland. So far there haven't been enough reburns at such short intervals to confirm this effect, though it has been reported in similar jack pines in northern Alberta.

Since 1994, there's been a major epidemic of at least one bark beetle species somewhere in the West every year. The biggest one, the mountain pine beetle epidemic of the 2000-aughts, peaked between 2006 (in B.C.) and 2010 (in the United States). In B.C. it is already on the rise again. In Colorado that species was back down to innocuous endemic levels by 2014, yielding its spot at the top of the state's insect damage charts to the spruce beetle from 2013 through at least 2017. The spruce beetle had first grabbed our attention in the 1990s with vast dead-tree panoramas on Alaska's Kenai peninsula, where most trees are spruce. In 2000 to 2003 the Southwest suffered extreme drought, leading to outbreaks of piñon ips in New Mexico and southern Colorado, and western pine beetle in the Southern California mountains. In the present decade, the Douglas-fir beetle has mounted an ominous increase in British Columbia and around Yellowstone. The Mexican pine beetle threatens to expand its range north from Mexico, and the little-known roundheaded pine beetle has hinted that it could become an epidemic species in the Southwest.

And then there's California, circa 2015.

The Sierra Nevada is an enormous fault block range. The east side of the range is steep, reflecting the fault that raised it, while the entire western slope is boringly gradual: from the first foothill to the crest of the range you'll log upward of eighty highway miles. Along the way you'll see an elegant display of changing forest zones, or belts, reflecting a gradient from hot and dry lower elevations to cool and wet heights. In gold prospecting days the lowest foothills were marked by the first few oaks interrupting grassland, but today the lowest foothills may support stone-fruit orchards. Above the orchards, gorgeous oak savannas take over, sprinkled with grazing cattle. Sporadically mixed in with blue oaks you may see your first pines—gray, or ghost pines, a lesser-known species whose foliage is distinctively sparse, often droopy, and, yes, ghostly gray. Gargantuan pine cones litter the ground underneath. If you dare to walk there, wear a hard-hat: these massive cones weigh from one to two pounds, and appear to have been the inspiration for the medieval mace. (Similar but even heavier cones grow on the Coulter pine, a close relative in the California Coast Range.)

By and by, you'll start to see ponderosa pines; farther up, additional conifers join in, comprising the mixed-conifer belt. Deeply grooved red fibrous bark on a trunk that tapers sharply from its base would be incense cedar. Similar bark and taper also characterize giant sequoia, but when you eventually reach the giant sequoias they'll make these cedars look twiggy. If you keep the high branches that overhang the highway in your field of vision, you'll notice when the sugar pines begin: you'll see huge cones making branch tips droop in graceful curves, as if on an oversize Christmas tree. Sugar pine cones are not the heaviest pine cones but they are the largest, ranging between one and two feet long.

In 2014 through 2017, something went badly wrong on these scenic routes. Half the big pine trees died. Photographers hunting for shock value had no trouble framing entire forested landscapes where it looked like *most* of the trees were dead, and scientists easily found plots where literally all of the good-sized ponderosa pines were dead. Freshly beetle-killed sugar pine and ponderosa pine turn a golden brown quite different from the red phase on lodgepole pine. In photos of beautiful blue lakes lined with summer cabins, it looks like lovely autumn foliage color. But every golden tree is dead.

Sugar pine is a magnificent tree, the largest of all pines. In his 2001 book, *Forest Giants of the Pacific Coast*, Bob Van Pelt described five rivals for the title of Champion Sugar Pine. All five are dead today. One of them was eleven feet six inches in diameter; six tall people encircling it were just able to touch each other's fingertips.

In random plots in Sequoia and Kings Canyon National Parks within the mixed-conifer belt, 70 percent of the big sugar pines were dead in 2016, and half of the big ponderosa pines. Other parts of California's ponderosa pine country were also hit hard.

I talked with Nate Stephenson, the U.S. Geological Survey forest ecologist who directed that mortality inventory in the parks. "Those are shocking numbers," he said. "That's a lot, but it's still a forest. From a distance, when they had just died and their crowns were all orange, since the pines are usually the biggest trees on the landscape, it was really dramatic." Mortality among the younger pines and the other mixed conifers—mainly white fir and incense cedar—was considerable, but below 25 percent. Those smaller survivors tend to hide behind golden pine foliage in the photos. A year later the forests looked greener: "Once they dropped all their needles and you look at the landscape, you'll see gray skeletons of pines standing around, but it still looks mostly green."

Incense cedar seems to be the species best poised to take up the slack— or at least the part of the slack that doesn't get eaten up by fires and then chaparral. The cedars were already a big part of the density infill in California mixed-conifer forests throughout the fire exclusion era, and perhaps for that reason alone California ecologists give them no respect. "They grow like weeds." But when pressed, these same scientists laugh and admit that it's a good tree, rating above average in pest resistance, disease resistance, fire resistance, growth rate, aesthetics, and even market value. It has to reach a good size before the fire resistance and the market value kick in, but then that's true of a lot of the other conifers, too.

Oaks of several species, both deciduous and evergreen, seemed invulnerable to this drought, and are likely to encroach on the ponderosa pine belt, starting from their own belt immediately downslope. Some gray pines may tag along with them.

Taking their cue from Nate and other forest ecologists, the media called it an epidemic of drought. From 2011 through 2015, the state's drought was extreme, the driest four-year stretch in the past thousand years, by some calculations, and hotter than those other dry stretches. Californians found it easy to believe that such a climate aberration could kill forests. In fact, the grim reaper had many tiny helpers: a multi-epidemic of bark beetles was there, under the bark, doing the dirty work. Several beetle species ganged up on California—sometimes even on individual trees. The top killers were western pine beetle on ponderosa, mountain pine beetle on sugar pine, Jeffrey pine beetle on Jeffrey pine, and fir engraver on white fir. The western pine beetle's infrequent previous outbreaks also came after prolonged droughts—in the 1930s in the Northwest, and in 2003 in Southern California. This latest outbreak was by far the worst in recorded history for western pine beetle.

"With luck," Nate Stephenson told me, "there are some resistant trees out there that will stay resistant to bark beetles. Maybe some of the natural enemies of these beetles will have a chance to catch up to the population explosion and limit the beetles, maybe not."

Foresters predicted this epidemic back in 1995; they generalized that western pine beetles thin any California ponderosa pine stand that exceeds certain density levels. They predicted steady endemic beetle mortality beginning at one density level, or an outbreak at a higher level. Most of these forests were around that higher level.

To figure out exactly how this beetle explosion correlated with the drought and tree density, Derek Young and his team crunched numbers. They divided the forested part of the state map into grid cells and then looked for correlations between a map cell's tree mortality level and its shortage of water based on local precipitation, snowpack, temperature, soil water capacity, and—critically—the density of all the plants competing for that water. The key factors for higher mortality were the four-year drought, higher temperatures than in previous droughts, and too many straws (trees) sharing the available drink.

Teams involving Greg Asner came up with similar answers from more direct measurement of canopy moisture, using imagery taken from Asner's

Dornier 228, a twin turboprop. HiFIS digital analysis of the light spectrum coming off the foliage can detect the level of moisture in the needles of the tree crown. Flying about seven thousand feet up, they had sharp enough resolution to look tree by tree, identify them to species, and correlate year-to-year reductions in canopy water content with the likelihood of dying in the drought. Again, the correlation was strong.

Bottom line: there is little disagreement that this multi-epidemic can be pinned on the hot drought. Human management likely also played a role via the usual pattern—fire exclusion leading to forests that are too dense and too homogeneous.

Craig Thomas has devoted decades to advocating for Sierra Nevada forests through his organization, Sierra Forest Legacy. Summing up his reaction, he told me,

It's really confounding. We haven't seen something of that magnitude before. It causes you to rock back on your heels, like, "O-kayyyyy, what are we going to do now?" A tragedy has forced a reckoning. We're going to have to reestablish fire, or it's over. If all the wood stays there and we're not incrementally burning some of that fuel, and maybe going in where it's less dense and planting; if it's all lying there over the years and then it burns again, then we will convert more of this landscape [from forest to brush] than anyone's going to want to see happen. I don't honestly think we're going to get very far if we don't get the natural process [of fire] back. If we do that, it won't be like it was in 1800, but we could hold on to this ecosystem that we all recognize as valuable and beautiful.

Scott Stephens, who codirects the Center for Fire Research at UC Berkeley, led a scientific report that backs up Thomas's fears. When the large dead pines have dried out and many of them have fallen, he calculates they will represent an unprecedented fuel load with a heightened likelihood of engendering mass fire effects. "A different and dangerous class of fire behaviors emerges at large scales and depends on the combination of high dead surface fuel loads and long burning times." He (with his coauthors, who include fire dynamics expert Mark Finney) pictures fires so ferocious that firefighters can't even approach them, fires that burn so hot they go beyond 100 percent tree mortality and into the realm where they consume the soil's organic matter and the root crowns and seeds that would normally regenerate post-fire plant communities. While that scenario remains speculative, it's pretty clear that firefighters won't be able to go into beetle-killed forests during a fire because of the high risk of dead trees falling on them; that fact alone will increase the extent of some fires.

As the pine death toll mounted, countless Californians chimed in on forums with opinions about how the dead trees need to all be logged before they turn into a tinderbox. In fact, the number of big dead trees is way beyond California's capacity to log. There aren't enough logging crews or enough sawmills—especially sawmills ready to handle big trees. Working at full capacity, the available crews and mills have been able to go through just the "hazard trees," the ones poised to fall on roads or buildings. Ecologists I talked to expressed frustration at the industry's lack of interest in even trying to market blue-stained pine, which beetle-killed pine trees turn into after a year or two. By 2018, there wasn't a lot left that the industry wanted. So what can be done to ameliorate the fire risk? Essentially, the only option is prescribed burning—a lot of it—in spring, while the wood is too moist to fuel mass fire effects. A lot of it. A few carefully chosen thinning projects, where they're most practical, may help.

The one-two punch of beetles and fires got the attention, at last, of California's governor and legislature. In 2018 they authorized unprecedented sums of money for proactive management including forest thinning and prescribed fire. Cal Fire, the state agency whose mission had been fire suppression almost exclusively for many years, is hiring seventy-five specialists in prescribed burning. These are state funds, and will flow mainly to state and private lands, but there are ways for the Forest Service to partner on some of the work. It is a huge turnaround, at least in one state. Forest service budgets remain atrociously strangled. In 2020 when a new approach to funding firefighting kicks in, the Forest Service will be able to plan a budget for the whole year without the second half's budget suddenly getting sucked away for firefighting. But it will still be short of the level of funding that is needed.

Countless people over recent decades have used the word "tinderbox" to describe beetle-killed forests. So far, events have not unequivocally borne out that description, at least not in the United States. Fire scientists conducted dozens of studies attempting to confirm or to refute the alleged tinderbox, and there's no wide agreement among them. Some studies of individual fires found a tinderbox effect; one fire in central Oregon spawned two studies that came to opposite conclusions. But as more post-beetle fires came and went, producing bigger data, the momentum shifted. Studies analyzing the data for broad effects of beetle-killed trees on either fire extent or fire severity are tending to show no net effect. (These broad studies mostly look at the spectral reflectance of vegetation as recorded in satellite imagery. It's a relatively objective way to evaluate fire effects on vegetation across wide areas, and it has become the standard way to evaluate fire severity.)

Firefighters, though, who look at fires up close and personal, consistently report heightened fire behavior when fires move into beetle kill. It can get terrifying.

Scott Stephens's group looks at the enormous buildup of fuels and can't imagine that they won't have an effect. Their paper hints that the broad scale of remote-imagery studies may cause them to miss the beetle effects.

I talked with Garrett Meigs, the lead author of the study they were talking about. He conceded that "if there's any place to find a signal [of beetle kill exacerbating fire] I think it would be British Columbia." Bingo. No beetle/fire overlaps in the lower forty-eight states come close to B.C.'s six million acres of fire in 2017 and 2018, which burned mostly within B.C.'s sixteen million acres of heavy beetle mortality.

Sure enough, I hear a dramatically different view of the beetle tinderbox effect when I talk to Canadians: "Observations made by fire behavior analysts indicate that the fires that burned in the 2017 season were larger, spread faster, and burned more intensely than they would have in the absence of the beetle-affected fuels." That's from a report presented in 2018. One of its authors, Daniel Perrakis, wrote a paper based on an earlier fire; it remains the only tinderbox study based on actually looking at and measuring wildfires in progress. It sees clear effects—again, in terms of intensity and rate of spread. What about fire severity—the fire's lethality to plants? "We weren't remotely interested in fire severity!" exclaimed Perrakis. "That's not what will burn over a town, or not, or cause a firefighter to seek shelter."

The ideal study would set up and measure research plots with varying stages of beetle mortality, and then see them get hit by a wildfire. That's next to impossible to arrange. In British Columbia it would be hard to even find comparable lodgepole pine stands with no beetle mortality. And then you would have to get lucky, to get a good wildfire through your plots.

In Kootenay National Park, managers tried the next best thing: they set up plots with varying beetle mortality from the 1980s, then put heat-protected cameras and thermometers in the woods and lit a prescribed fire. In the highest beetle mortality area, the measured intensity and impact of the fire were anywhere from two to eleven times greater than models predict for that forest type without unusual mortality.

Currently Perrakis is using satellite data to look at behavior of the fires of 2017 and has found that beetle kill in B.C. definitely increased the probability of burning. He tells me a couple of reasons he thinks beetled fuel made the fires bigger than they would have been. One comes from a close-up observation of the affected fires. "They don't necessarily go in a downwind direction. Some would go sideways, to where the red-stage fuels were. We call it 'fuel-seeking.' Short bursts of independent crown fire." The likelihood of burning, for these patches, increased because they were dead and their needles had dried out and turned red.

A second reason is simply the big obvious correlation. "The weather is not breaking any records, and yet we are breaking fire behavior records. The year 1998 broke several B.C. temperature records that have still not been broken. There were about 2,600 fires that year, they burned about 185,000 acres. Why in 2018 did we burn sixteen times that? It certainly looks to me like beetles were an important factor."

Garrett Meigs, who sees little clear evidence of a tinderbox effect (in the United States), argues that the correlation between beetle mortality and big fires more likely stems from the fact that both of them are exacerbated by warming climate. Daniel Perrakis knows that argument well, but feels that the abrupt timing of the fire increase matches beetle mortality better than it matches either warming or long-term effects of fire exclusion. But he concedes that weather was a factor: even though heat and humidity for the year were not record-setting and wind speed was unremarkable, other measures may be more telling, such as days without rain, or days of unusually hot weather per month.

California had record-breaking fires in 2017 and 2018, but they do not correlate much at all with the map of dead trees from the peak beetle mortality years, 2015 and 2016. Remember, though, that Scott Stephens's warning concerned a phase that's still in the future for California, ten to twenty years after the mortality. Studies that model the changes in fuels following a beetle outbreak distinguish the "red stage" (dead needles still on the tree) from the "gray stage" (needles on the ground, trees still standing). They've produced a conventional wisdom that red stage is slightly more flammable than live forest, whereas gray stage may be less flammable. Some also describe a subsequent "old stage" (dead branches all on the ground, some dead trees fallen and jackstrawed, new tree saplings thick on the ground), which is again more flammable; but other studies haven't been able to look that far along, so there isn't as much of a conventional wisdom for the old stage. British Columbia's lodgepole pines were reaching the old stage in 2017, and they were well along in the old stage during the prescribed fire in Kootenay National Park. But it isn't just the old stage: Perrakis's study found higher intensities in both red and gray stages as well.

In sum, the jury is still out on the beetle tinderbox effect. Views from Canada and the United States diverge. Forest conditions vary enormously and interact with fire weather conditions in countless permutations. In at least one case in the Sierra Nevada, beetle mortality actually suppressed an intense fire. Dense whitebark pine groves were 100 percent killed by this fire, along with their understory plants, whereas among nearby white-barks that had once been similar but were now 60 percent beetle-killed and in the gray stage, the fuel was apparently too sparse to carry an intense fire.

Perhaps some north-versus-south generalization is real, or it may be more a matter of different studies using different methods, on different scales, at different fires, and asking different questions. Perrakis is interested in danger to firefighters and to communities; Meigs is more interested in ecological effects. Still, both are interested in the likelihood of burning, and their conclusions on that differ.

Scientific studies also duel on the question of whether we can reduce the beetle risk by reducing forest density. A majority again finds that it can; but several caveats are in order. Pine species vary and so do beetle species: the case is strongest for ponderosa pine, sugar pine, and Jeffrey pine, and practically nonexistent for lodgepole pine during a *mountain* pine beetle epidemic. That's a big weakness, considering that the post-1998 mountain pine beetle epidemic in lodgepole is still the grizzly bear in the room. At times, hopes were raised that the attack could be held at bay with thinning focused on removing infested trees. Those efforts are hardly ever cost-effective: once beetle numbers are so high that the flying adults swarm into new territory like clouds, they have no trouble overcoming the best-defended lodgepoles.

That's mountain pine beetles: when their populations explode they switch from seeking weakened hosts to seeking the healthiest ones they can find. In contrast, western pine beetles and Douglas-fir beetles may continue to seek stressed trees while erupting, so the tree species that they attack are much more likely to benefit if we reduce density and stress—before a western pine beetle outbreak happens.

To keep the issue clear, we have to separate the failure of sanitation thinning to stop epidemics in lodgepole pine from the partial success in mitigating epidemics of preventive thinning done *before* beetles erupt. The evidence is fairly good in pines other than lodgepole. And remember that there's little or no advocacy anyway (at least among experts) for either thinning or prescribed fire in lodgepole pine, nor in spruces and firs.

The simplest reason why thinning may work is that with fewer trees drawing water out of the soil, there's more water for each tree. Better-watered trees have more resources to devote to their beetle defenses, which consist mainly of toxic pitch. Drought stresses the trees, starving their defenses; thinning ameliorates drought. (Density reduction—either by beetles or thinning or fire—often has a cool side effect: it leaves more water not only in the trees but also in the streams. See chapter 2.)

"Fewer straws in the drink" is just one possible reason why thinning trees may reduce bark beetle attacks. Other, trickier mechanisms have been studied: the number of potential host trees that are close enough together for the smell of beetle pheromones to carry from one to the next; wind speed between the tree trunks; the effect of shade on the temperature of tree bark. Effects of each of these have been confirmed in one place or another, but they're unlikely to be major factors across the whole West.

Another intriguing possibility is a direct effect of fire on pitch production in ponderosa pine. Sharon Hood, of the Forest Service's Fire Sciences Lab in Missoula, thinks that frequent low-severity fire, back when that regime was still in charge, served to maintain ponderosas' defenses by inducing growth of more pitch ducts—and would so serve once again, if we bring fire back. It's been noticed almost forever that old ponderosas that survived numerous fires have pitch-saturated wood near their bases. Studies have also found pitch flow increasing soon after a fire. What's harder is connecting that effect to increased survival under attack from beetles. Many pines that survive a fire die from beetle attacks over the next few years, because the fire injuries attract beetles. This delayed mortality overwhelms any short-term improvement in defense that may result from pitch flow. It seems that if Hood's "inoculation" hypothesis is correct, it may show up only after there's been a whole series of light fires, like in the old days.

Hood has found the pitch-duct increases she's looking for in certain studies—hers as well as some other scientists'—but in other studies it doesn't show up. Studies of particular fuel treatments and their effects tend to be confounded by innumerable variables. Each one may suggest a new wrinkle. For example, Hood's latest one found increased resin flow after thinning but *not* after burning without thinning. Was that burn too intense? Not intense enough? And yet, burning alone did yield a significant improvement in survival through a beetle outbreak. Hood proposes that it may have enhanced the toxic chemistry of the pitch while for some reason not increasing its flow rate or the number of resin ducts. A few studies by other researchers found the reverse: distinctly increased resin flow after fire, without any improvement in resistance to beetles.

I think we may have to wait awhile to see this hypothesis either vindicated or not. Evidence one way or the other should accumulate pretty quickly, as increasing numbers of fires touch more and more ponderosa pines. The beetles aren't going away. One hopeful anecdote comes from the parts of Yosemite National Park that have enjoyed a restored fire regime for forty-seven years: they experienced far lower beetle mortality as of 2016 than similar forests elsewhere in the Sierra Nevada.

Early in this chapter I allowed that bark beetles belong here—part of the natural order. But could their epidemics in any way be a good thing, the way forest fires are often a good thing?

Some people see some benefits. I'm not just talking about incorrigible fans of all things six-legged. First off, we can generalize that tree species or geographic races are not going to be able to migrate north and upslope as fast as their preferred climate does. Facilitating their migration is a good thing. Most of them are mainly able to move in where space is cleared for them by disturbances of the existing forests. Fire, pests, and logging can all perform that service, even though all also carry a risk of converting forest to nonforest vegetation under some circumstances.

Secondly, lodgepole pine forests and subalpine spruce-fir forests tend to grow uniformly old, dense, and arguably stagnant, with nothing growing on the forest floor, just lots of brittle dry twigs and branches and needles, because the trees are sucking up all the available water. You can make a case that pest epidemics, like fires (even though the fires there tend to burn at high severity), renew these forests, allowing youth and diversity to flourish anew—allowing flowers to bloom, literally.

It may actually be better than that. On the high and dry east slope of the highest Sierra Nevada grow stands of limber pine that suffered heavy beetle mortality in the drought of 1984–1992. (At the time, that was the worst and hottest California drought in a century, but the drought of 2012–2016 outdid it.) These were dense "young" pole forests of trees averaging barely two hundred years old, in contrast to sparse limber pine old-growth not far away, where many pines are well over one thousand years old. The years 1999 to 2004 brought another drought, but that one caused no further mortality in these stands that had been nicely thinned out by the beetles a few years earlier. Though the increase in their drought resistance could be a simple effect of thinning, there are clues that it was more than that.

Connie Millar, a Forest Service geneticist, decided to compare their growth histories. She cored a few hundred live and dead limber pines and found a striking pattern: the dead ones (killed by beetles) had been faster-growing than the live ones back during the Little Ice Age of the 1800s, but in the 1920s the growth advantage switched: now, in the warmer twentieth-century climate, the live trees (survivors of the recent beetle epidemic) started growing faster than the others. The correlation suggests that a small leap in the evolution of limber pines had just happened here; the population better adapted to a warmer climate had been selected, and the maladapted genes were swept away. This fits well with general evolution theory: evolution proceeds by little fits and starts, and they typically occur during mass mortality events.

This was exciting stuff, especially if it was happening broadly. So Millar pursued it, studying another kind of pine in the same region—whitebark pine—and she found the exact same pattern. "Improved fitness through forest dieback" is one phrase Millar uses for her hypothesis. When I asked her about it she was unabashedly enthusiastic. Indeed, she takes a strikingly benign view of beetle epidemics: they do our forest thinning for us, and generally help forest trees adapt, by both culling trees with worse genes and making space for growth of trees with better genes. But when I asked what she has come up with since 2012, I found she has moved on to other climate-related topics. (Pikas; limber pines leapfrogging past bristlecone pines in upward migration, as described in chapter 11. Millar started out as a forest geneticist back when that meant breeding faster-growing trees, but she has branched out considerably.) As for the progress of her hypothesis since 2012, she told me "there is one person who has picked it up and run with it. Whatever she tells you will be the latest word. Diana Six."

Diana Six teaches and does research at the University of Montana in Missoula. In graduate school, she landed a post that combined bugs and fungi, two things she is inordinately fond of. She embarked on a specialization in the symbioses between bark beetles and the wood-rotting fungi they carry with them. She's also fond of beer. Putting the three together, she took fungi from a bark beetle's mycangium—the crevice on its head that's specialized for transporting fungal cultures—and she cultured them on petri dishes, purifying again and again until she had pure cultures of the type of fungi we call yeasts, and made beer with them, three times with three different yeasts. One turned out pretty okay. She's more enthusiastic about bark beetles as snacks, and gets her students to try them. They taste a little like pine nuts—not surprisingly, since the same pine oils enrich both.

In her Clark's-nutcracker-colored Mini Cooper messily crammed with field gear and papers, we set off for a day's loop of Montana mountain ranges—Flint Creeks, Anacondas, Pioneers, Beaverheads, Bitterroots, Sapphires. I'm convinced it's mountain ranges that make the Montana sky a Big Sky—so many distinct, well-spaced mountain ranges, each arranging its own afternoon clouds, and all dwarfed by the great blue dome.

Our route up Interstate 90 parallels the Clark Fork and a busy BNSF track bearing long trains rumbling westward to feed the world's fossil fuel hunger. Trains laden with Powder River coal alternate with trains full of Bakken oil.

On the day of our Big Sky loop, Six has recently learned she is to receive the Edward O. Wilson Biodiversity Technology Pioneer Award. Previous winners include such gods of twentieth-century science as Lynn Margulis, Alan Turing, and Tim Berners-Lee, who "invented" the internet.

Her career arc is all the more astonishing for having caught up after a rocky start. Dad was a "hillbilly" with a fifth-grade education. Mom was a German woman he met while stationed overseas. Mom's German surname, Six, was said to descend all the way from the Roman General Sextus Julius Frontinus, who governed the Rhineland in the late first century. The couple "had nothing in common. He was country, she was classical. They fought all the time." The one good thing about their child-rearing was that they took Diana and her sister camping every weekend in either mountains or desert not too far from home, east of Los Angeles. But Diana fled the household at high school age and joined a biker gang. She married twice in succession, two abusive marriages. "They're both dead now. (It wasn't me!) I almost died a couple of times. Somehow I figured out I had to change something. But, you know, to leave a biker gang you have to disappear."

Once her disappearance was accomplished, she found her way to community college for a GED after nine gap years. "There was a lot besides academics. I had to learn to talk. Before, I couldn't put three words together without the f-bomb. I wore black leather, I scared people."

She had figured a high school diploma would do it for her, but those pesky teachers kept goading her to continue, so she started community college. In those days it was free in California. No free tuition, no Dr. Six. "I didn't have a clue. I majored in library science because I knew I liked to read." Gradually the interest in bugs and fungi rose like cream to the top, and teachers kept nudging her along. "I got my PhD at forty. Now I'm sixty-two, and I'm just getting started!" She says she works 24/7, but I also hear about all-day bike rides across the Big Hole to Wisdom, Montana, and about fly-fishing the Wise River, so she fits those in somewhere, along with brewing bark beetle beer. Oh, yes, she's also pursuing a master's in journalism, to improve her ability to communicate to nonscientists.

I'd rate it well above the average entomologist already.

Just past the Continental Divide signs we turn west up the Big Hole River and then south into the Pioneer Mountains. In 2010 the Pioneers were "an entomologist's paradise." Six would ascend through Douglas-fir forest infested with western spruce budworm (massed fluttering moths, actually, whose caterpillars are the so-called worms) and up into the zone of lodgepole and limber pines in full mountain-pine-beetle-epidemic mode, clouds of tiny flying beetles.

Two years earlier a colleague had detected some of the epidemic's first little scouts in the Pioneers. They started out up high, on whitebark pines. It was a perfect opportunity to scientifically observe an epidemic from its beginning. "When we started working out here it was a really exciting experience to see it just start to take off. A couple of years later," when they saw the sheer scale of pine death, "it moved from excitement to just heart-wrenching. We moved into a state of depression for the last few years working here. Then I saw that there were survivors, after all. It changed my depression around to making me feel like maybe there's some hope."

We walk through a forest of dead whitebark pines. The beetles are long gone. She pulls a black titanium hatchet from her pack and flakes some bark from a tree that looks like a relatively fresh kill. A few desiccated pupae remain. The next tree was killed by twig beetles, a sort of miniature bark beetle—a pain to study because they're hard to see. If western pine beetle larval tunnels look like spaghetti, a twig beetle's tunnels are vermicelli. Until very recent years, twig beetles were mostly confined to twigs, and weren't seen as a deadly threat. Now they are killing pines. Little is known about them.

Now that the dead trees have all lost their needles, green survivors appear more prevalent in the stand. "There were big mature trees that looked to be perfect hosts for the beetles, but the beetles didn't kill them. In fact we couldn't even find any attempted hits. That didn't make any sense to me. Why would they just ignore perfectly good hosts?" At the peak of the population boom there was plenty of pressure on the beetles to exploit every possible food source, and they came so close, yet spared these trees. Why?

She compared the trees' DNA using the molecular methods taxonomists use to sort out genetic relationships. Killed and living trees fell into two distinct lineages, confirming that the beetle epidemic was a natural selection event, shifting the genetics of whitebark pine populations—at least within the many hard-hit locales. What trait was selected for? Resistance to mountain pine beetles is the logical one to start with. From that much alone, Six draws a lesson for forest managers: if the survivor trees stick around to provide most of the seeds for the next forest, they will improve its genetics. But if you clear-cut the area after the epidemic, you're defeating natural selection and assuring that the next crop of pines will be just as beetle-prone as the last one. And that practice has been all too common, especially in British Columbia. It's understandable, because unattacked trees are worth more at the mill, and loggers and political leaders alike want to come out in the black at the end of the operation. But it's shooting ourselves in the foot—or perhaps I should say shooting our grandchildren in their feet.

Wanting to find out what additional traits may have been selected, she looked for a chemical difference between the two groups. She expected to find higher levels of defensive terpenes in the survivors. Stunningly, the results were just the opposite: somewhat lower terpenes in the survivors. To make that trend meaningful, the most plausible explanation would derive from the fact that pine terpenes, aside from being toxic to beetles, are also the aromas beetles use to locate pine hosts. The terpene we think they rely on most, alpha-pinene, was especially low in the surviving trees. "It may be that the beetles just can't find these trees."

She's pleased that she can demonstrate a clear genetic difference between beetle victims and survivors. The survivors are genetically superior, in that they resist beetles. But when it came to the part of Connie Millar's hypothesis about survivors being a race better adapted for a warmer climate, Dr. Six's studies could not confirm that: this time the survivors were the slower-growing whitebarks, not the faster-growing ones. Other scientists have submitted evidence that slower growth may actually be a drought-resistant trait, and that does make sense. It matches the pattern of piñon pines that survived the last beetle epidemic in New Mexico, as well as the fact that bristlecone pines are among the slowest-growing, most beetle-resistant, and most drought-resistant of western pine species. "The important point," to Millar, "is that there is selection for greater fitness as a result of bark beetle attack. The nature of the fitness-related traits no doubt differs" between the Sierra Nevada and the Montana Rockies.

Jeffrey Pine

4 COOKIE CUTTERS

A few miles south of the U.S.–Canada border stands a mountain . . . well, really little more than a ridge, by the standards of its neighbors. (They're sometimes called the American Alps.) It's so remote that, though you can see tens of miles from there in any direction, at night you can't see a single artificial light except for passing jetliners and the occasional truck rounding a bend in a small highway twelve miles away as the raven flies.

In 1926 a fire swept through the mountain's forests, leaving a desolate scene for which, later that year, the mountain was named Desolation Peak. In the decades after the fire, plants did what they usually do, rendering Desolation a misnomer today. We could call it Inspiration Peak, if that name weren't already taken by a sliver of a spire twenty miles to the west. In 1932 the Forest Service built on it a cubical pyramidal rustic hut walled with eighty panes of glass: a fire lookout. The view is, as they say, commanding: in most directions, a steep 4,400-foot drop-off to the glaciated trough of the Skagit River, countless active glaciers on the peaks rising abruptly from the valley's far side. Between 1950 and 1954, the valley floor turned into a deep-blue fjord lake a mile wide and twenty-three miles long, as the Skagit's waters rose behind a new dam. In 1956 a rather dissolute unsuccessful writer manned the lookout. Moved by the view—"brilliant and bleak beyond words"—and by his loneliness, he later (after achieving success) published the journal he wrote there as the first section of a novel, *Desolation Angels*. Its opening pages offer an ode to his dark angel, Hozomeen, the mountain adjacent to the north.

Hozomeen means "knife-sharp." Though hardly a household word, Hozomeen does catch the attention of certain mountain data nerds who analyze mountains to rank them, looking for maximal vertical relief in minimal horizontal distance. When all contour lines were crunched, Hozomeen turned out to be statistically the most impressive nonvolcanic peak within the ruggedest mountain range within the lower forty-eight states— Washington's North Cascades. The writer/lookout, whose name was Jack Kerouac, didn't know that. To him, Hozomeen was "an imperturbable surl for cloudburst mist," it was "the Void."

Thirty-three years later, another barely published writer manned the lookout. Shortly after his arrival, several tons of Skagit gneiss fell onto the North Cascades Highway, blocking it for weeks, and visitors to Desolation dwindled to a desultory trickle. The few who made it, aside from the lookout's bride on a brief visit, were mostly Kerouac pilgrims. (Kerouac's sixty-three days there were of course before the days of Kerouac pilgrims, and his visitor count rang up as a big zero.)

One morning—too early in the morning for a visitor to have hiked from the lake, so they must have camped at the little-used campsite a mile from the summit—I had to tear my eyes away from Hozomeen to see who was hallooing at the door. It was a forestry graduate student, drawn to Desolation not by its literary history but by its fire history. She was scouting out locations for a tree-ring study. Since my ridgetop was generously sprinkled with ripe blueberries of a species aptly named *deliciosum*, and I had a bit of butter, flour, and sugar and plenty of time on my hands, I invited her to come back that evening for blueberry cobbler. The lookout cabin in 1989 had a small propane oven and a propane mini-fridge which did nothing to tarnish the ambient soundtrack. (Little-known but wondrous fact about propane refrigerators: they are silent.) A helicopter brought full propane tanks every couple of years.

Decades pass. The visitor, Emily Heyerdahl, has made a name for herself with outstanding work with tree rings, I've written a few books, and though we were essentially strangers, we were simultaneously looking for each other amid the scientists thronging the lobby of a wildfire science congress. I had questions in her area of expertise; she wanted to ask if I remembered baking her a fire lookout's cobbler.

Desolation didn't make the cut, she told me. Too many fires in its past meant not enough old trees left. She picked Crater Mountain, coincidentally the next fire lookout peak to the south, the one where the poet Gary Snyder spent the summer of 1951, inspiring Kerouac to try it five years later. Gary liked the solitude, Jack didn't. Anyway, Crater Mountain hosted Heyerdahl's first big fire scar study.

Heyerdahl's relationship with wildfire is conflicted. She works out of the Rocky Mountain Fire Sciences Lab in Missoula, Montana, and loves her work. (In fact, the receptionist there told me it's her favorite workplace ever, because everyone there loves their work.) But in recent years, her asthma and Montana's smoke conspire to chase her out of the entire region for most of the summer, starting before the first summer day that even thinks about dry lightning. Now even her prized collection of tree sections and cores is fleeing Missoula, headed for a climate-controlled tree-ring archive in Tucson.

Tree-ring research has several branches. The central one, dendrochronology, looks at the tree's annual rings all the way to the pith, or center. That tells you how old the tree is, of course, or how old it was when it died. Close study of the sequence of thicker and thinner rings tells you a lot more. When a tree grows faster, it produces a wider annual ring. Faster growth, if shared broadly across a region, indicates a climate trend favoring tree growth. More rain at the right time of year, for example. A cooler summer, if it's in one of the many parts of the West where trees are drought-limited, or a warmer summer, in a high mountain area where trees are limited by cold. An abrupt switch from many years of thinner rings to several steady years of wider ones suggests the tree was released from competition because one or more of its neighbors died. If lots of trees near each other began growing faster in the same year, it suggests they survived a fire or a disease or an insect outbreak that thinned out other trees from the forest.

On the other hand, a several-year period of relatively thin rings shared among numerous trees in an area could mean they themselves suffered an outbreak of defoliating insects—like the western spruce budworm, if the pattern is shared among the species that the western spruce budworm attacks. Any sequence that matches up across all the tree species of a region most likely reveals a swing in the climate.

Since climate cycles are regional, long-term sequences of thicker and thinner rings among nearly all trees of broad regions share identifiable patterns, even though no two trees are exactly alike. Dendrochronologists learn the patterns and can spot them immediately, like knowing Morse code.

In arid country, a dead tree can stand or lie around for centuries—a few last for millennia—without rotting. A dendrochronologist can section it, look at its pattern of rings, and tell you which century it lived in. This "cross-dating" of long-dead trees can piggyback onto the tree-ring record found in live trees. Of course, a giant sequoia forest provides a three-thousand-year record just in its living trees or its fresh stumps alone.

Tree rings provide the best record we have of climate fluctuations within the last few thousand years. For example, they tell the role that drought played in depopulating Mesa Verde cliff dwellings and the surrounding area in the thirteenth century.

Emily Heyerdahl pursues another branch of tree-ring research: fire scars. Dendrochronologists can do their work based on cores slenderer than a pencil; they also sometimes use entire "cookie" cross-sections, mainly of trees already dead. Fire scar researchers use a partial cross-section cut through a fire scar called a catface. "Why they're called catfaces," she tells me, "is a mystery." A catface is a patch of tree trunk, usually charred black, typically a triangle with its base at the tree's base. With two horizontal cuts of a chainsaw two to four inches apart, she takes out a section that includes the entire catface and a few inches farther into the tree. This type of cookie doesn't need the tree's center; if she needs the tree's whole life story she'll take a core. Since this side of the tree was cambium-dead already, cutting a cookie from one side doesn't unduly risk the tree's life. She tests that hypothesis every five years by revisiting one of her first study sites to check for mortality, and reports back to the research community. Over the first twenty years, the partial-cookie-cut trees were not dying significantly faster than similar trees in the controls.

The catface originated when a surface fire heated the bark enough to kill the underlying cambium—the living, growing, sap-carrying layer right under the bark. If cambium is killed all the way around a tree, the tree dies, but trees regularly survive cambium death up to 25 percent of the circumference. (Sometimes much more: see chapter 11.) Within a few years the bark falls off the dead cambium, exposing bare wood. Subsequent surface fires char the bare wood. As the tree grows a new cambium layer (and annual ring) each year, the edge of that new growth forms a rounded rim around the charred scar. A section sliced through the catface shows all of those rounded rims—an irrefutable graphic history of fires that that side of that tree has been touched by, complete with the number of years in between. It actually graphs a *minimum* set of fires, since other fires likely passed nearby without scarring this tree.

Early foresters in the West suspected that the classic parklike open stands of ponderosa pine were produced by frequent low-severity fires. Fire scar researchers confirmed it, finding fire intervals ranging from three years to around twenty years at different frequent-fire stands throughout the West.

Back at the Fire Lab, the cookie is recut with a bandsaw to get it smoother, then sanded with progressively finer grit until it is so smooth that the researcher can look at individual cells, and can easily distinguish a spring fire from a fall fire by where the fire scar hits within the space of one annual ring.

Since these trees with catfaces survived fire after fire after fire, the kind of fires they record necessarily tend to be low-severity fires—the kind that leave numerous survivors. That presents an incomplete picture. It fails to record the patches of highly lethal fire.

Most low-severity surface fires have some unburnt islands and some flare-ups that kill a patch of big trees. Toward the other end of the scale, the big 1988 Yellowstone fires were generally severe, but they were also patchy, with many unburnt islands.

All that patchiness is utterly vital to our forests. Patchiness of fires creates heterogeneity in the forest that helps to limit outbreaks of insects and disease, and also breaks up subsequent fires, limiting their size and making them patchy all over again. The heterogeneity persists forty years, a hundred years, even four hundred years. Patches of surviving trees are often close enough to provide seeds for regeneration within the 100-percent-mortality patches. (Most species do need survivors nearby, but a few of our pines are dramatic exceptions, including lodgepole pine, whose seeds rain down from dead trees, and whitebark pine, whose seeds fly for miles and miles with avian help.)

Emily Heyerdahl's contribution to the field was to tease out the record of the medium- to high-severity patches. High severity is shown by a cohort of trees all the same age; they date from shortly after the last fire, which killed all of their predecessors. Heyerdahl showed that if she very carefully and exhaustively combined fire scar sampling with lots of age sampling of trees lacking fire scars, she could identify subtle cohorts intermixed with older survivor trees, and use them to draw a map of a mixed fire history. She calls the ponderosa pine–dominated forest "a mixed-severity fire regime dominated by low-severity fires." (It's always been clear that forest fires tend to be patchy. Saying "low-severity regime" makes it sound like a dictatorship where sublethal flames rule absolutely. That's misleading. Many fire ecologists prefer to describe most western forests as subject to mixed severity rather than low or high. That's certainly more accurate; but others rightly complain that if nearly all forests fall in the "mixed" bin, the term is too broad to be useful. To avoid entanglement with low versus mixed, some prefer the term "frequent-fire forests." The ecologists focusing on a better way to discriminate among types of fires say the key is in the size and distribution of patches of the two extremes on the scale—high severity, and unburned islands. High severity is a long-term problem for forests mainly when the high-severity patches are so large that substantial areas are far from any surviving trees that can supply seeds.)

Andrew Merschel was a graduate student when Emily Heyerdahl worked with him (and coauthor Tom Spies) on a study that helped resolve a stubborn riddle about western pine forests: exactly how much have they changed since pioneer days? Merschel, with green eyes, black glasses, and a telegenic smile, showed me around a small, intensely piney-aromatic room of bookshelves crammed with pine and cedar sections on the Oregon State University campus. Since the sanding is done in another building, the biggest machine in this room is a specialized microscope with its head on a long rail so that it can glide back and forth over the biggest cross-sections. The magnified image fills a computer screen where he can click on each ring. The computer instantly measures the distances from click to click, analyzes their patterns, matches them up against a huge database (like fingerprint matching), and recognizes the decade, confirming what Merschel already knows, since he has learned tree-ring code. He studied the zen of western tree rings under Heyerdahl's tutelage in Missoula.

Merschel grew up in the very highest-elevation part of ponderosa pine range—Twin Lakes, Colorado, elevation 9,200 feet, where his family ran the Twin Lakes General Store and Post Office. His study area, the eastern, or dry, side of Oregon, has a lot of sagebrush, but also a lot of mountains high and cool enough for forests including ponderosa pine to mantle them. (Mountains are almost always a wetter habitat than nearby lowlands, partly because mountains provoke clouds and rain, and partly because they have cooler temperatures, which reduce the rate of moisture evaporation from both soil and leaves.) The driest margins of eastern Oregon forestland, often just upslope from sagebrush steppe or juniper woodland, support many pure ponderosa pine stands—some of them like the beautiful "pine openings" on the Oregon Trail. These sites are slightly too dry for most of the other conifer species.

Just one tick wetter than that is the mixed-conifer forest, the most widespread forest type in eastern Oregon. Typically the older trees in it are ponderosa pine, the younger ones are grand fir* and Douglas-fir, and other species contribute modest numbers. The seedlings and saplings are almost all fir, because in the absence of fire most stands have become too shady for baby ponderosa pines; grand fir seems to be taking over. Absence of fire has become the twentieth-century norm, even though here and in the Rockies (and in contrast to northwestern Oregon) summers bring plenty of lightning to start fires.

Merschel and Heyerdahl wanted to figure out what today's mixed-conifer sites were like in 1890 in two mountain ranges of eastern Oregon. They cored more than four thousand trees, including old stumps, in 180 study plots. Their results hit me in the face like a splash of glacial stream water. In 1890 there just wasn't much mixed-conifer forest. Most of the Douglas-fir and grand fir grew after Euro-American settlement, in what had been ponderosa pine forests up until then. Almost all of the large, old trees on those plots were ponderosa pine, regardless of how much fir grows on the plot today. There were plenty of firs around, but relatively few were reaching old age back then, other than on the especially cool and moist sites. Fires were coming along every ten or twenty years and killing most of the young

---

* In calling these trees grand firs (*Abies grandis*) I follow the guidance of Oregon's taxonomic authority, the Oregon Flora Project. Elsewhere in this book you'll find white fir (*Abies concolor*) described as playing the same role I'm here describing for grand fir. White fir is a close relative that abounds from California to Colorado to New Mexico and Arizona. Where their ranges meet (including Oregon), the two species hybridize, or intergrade. These hybrids in eastern Oregon have often been called white firs. However, the Oregon Flora Project determined that Oregon has no pure white firs, only grand firs and hybrids, so it does not list white fir as an Oregon species. It's simplest for me to just call the Oregon populations grand fir. Both species are "true firs" as opposed to Douglas-fir, a different genus; but if I say just "firs," I'm speaking of true firs and Douglas-fir collectively.

firs. Of course there were patches of higher severity fire, producing post-fire cohorts that may well have included all three species, but subsequent fires culled most of the firs.

Merschel points to other studies finding that grand fir has little chance of ever being a pillar of extensive healthy mature forests here. (This applies to the drier two-thirds of Oregon's mixed-conifer forests.) It grows fast and well for decades, but sooner or later an extended drought comes along and it gets wiped out by root rot, beetles, or spruce budworm. Count on climate chaos to hasten that fate.

Bottom line: over much of the region, removal of grand firs from the mixed-conifer forest can be honestly justified on grounds not only of forest resilience but also forest "naturalness." The grand firs' abundance in these forests stems from three things: first, unnatural fire exclusion; second, logging of the biggest pines, making space for firs; and third, the late nineteenth century being cooler and wetter than the twentieth century—let alone the twenty-first century.

Studies that pointed out true firs' health disadvantages would often extol the superiority of ponderosa pines. Now we've been reminded that ponderosa pine forests are also vulnerable to pest devastation. That said, removing the grand firs from a mixed-conifer forest would leave a sparser forest dominated by pines—closer to the nineteenth-century model and more resilient to twenty-first-century threats.

The Laboratory of Tree Ring Research in Tucson, Arizona, which now houses Heyerdahl's cookies along with many, many others, was long the domain of Tom Swetnam—the Tree Ring God, some call him. He recently retired to a home a scant eight miles from Las Conchas, New Mexico, where I meet him—a friendly bear of a man with a massive handshake, a white beard, wisps of hair, and a tiny, fluffy white dog of some thirteen years' standing. Many tree-ring scientists were his students. In his own eyes, he merely inherited the mantle of the original dendrochronologist, Andrew Ellicott Douglass, both in general and at the Tucson tree-ring lab in particular.

An astronomer, Douglass wanted to prove that sunspot cycles affect the weather. It first occurred to him in 1901 while descending from the Kaibab Plateau in a horse-drawn wagon that tree-ring widths might record climate cycles based on solar cycles. Eventually the American Museum of Natural History heard of his new science, dendrochronology, and engaged him in solving an archaeological puzzle: how old were the Puebloan ruins like Acoma, Chaco Canyon, and Mesa Verde? He sectioned vigas (beams) from old pueblos and crossdated them, establishing their relative ages, but he struggled for decades to complete a time line continuous to the present day. He went to California in 1915 through 1919, cut into giant sequoia stumps with a seven-foot crosscut saw (which looks ridiculously small next to those stumps) hand-sanded the sections, and came up with the trees' complete life stories.

Disappointingly, their climate was too different from New Mexico's to let him crossdate vigas. One bristlecone pine alone could have done the trick, or even certain old New Mexico Douglas-firs, but this was the dawn of tree-ring study and no one knew yet how to recognize an exceptionally old tree. Finally, in 1929, archaeologists pulled a big viga from a ruin right in front of him; he sectioned it, studied its rings for hours, and spotted pattern matches both to older vigas and to living trees. He then began rattling off the ages of dozens of pueblos to the astounded archaeologists. That beam was the bridge across time. It made him famous.

That's one of many stories I hear from Swetnam. His own recent story is that he retired to a woody octagonal home on an airy perch overlooking the valley where he grew up the son of the district forest ranger. He now chooses to become one of those WUI ("wooey"; wildland-urban interface) settlers whom he often admonishes about fire risk. It gives him a chance to show how to do it right.

Retirement from academia leaves him with time to pursue scientific tangents he couldn't indulge in before. He's trying to grow corn, beans, and squash here with different levels of irrigation, to explore the challenges to growing the traditional crops on that site. So far, it's convincing him that the Indigenous people were wizards at farming and had to have worked out some form of irrigation here. He also has a cool Riga greenhouse made from a kit, and he maintains three solar-powered weather stations whose data I can pull up on a weather app on my phone.

But the pursuit that he feels retirement has really liberated him to spend time on is . . . you guessed it: counting tree rings. Working in his personal tree-ring lab, a ten-by-six-foot outbuilding behind his home, he prefers to match ring patterns the old way, rather than leaving it to a computer. It's called a skeleton plot: he slides a pair of long inch-wide strips of graph paper on which he has marked each year's width on one graph line. One strip is the tree in question; the other is the master plot carefully compiled from as many nearby tree cores as he can find.

Much of his recent work has been with the Jemez Pueblo, the people of this valley. Their ancestors lived here for hundreds of years, apparently coexisting with the fire regime—and that's where Swetnam finds at least an oblique refraction of reassurance: the Puebloans proved that it's possible to safely inhabit New Mexico pine forest.

(The word *pueblo*, originally "people" and later also "town" in Spanish, applies in several ways: Puebloan peoples lived in villages and sizable multistoried towns, called pueblos, built of stone. Today, some of them live in places called Pueblos that are also government entities, essentially small reservations. The word *Jemez* is a Spanish attempt to spell this local group's word for themselves, pronounced HAY-miss today. They live in the southern part of the Jemez Mountains, which are the broad volcanic shoulders of the Valles Caldera. The mountains grew through countless eruptions from many vents all drawing from a magma system that collapsed more than a million years ago to produce the broad round caldera. The caldera floor is at 8,000 feet and its rim goes up to 11,561 feet.)

The Jemez population before 1650, just in this drainage alone, was somewhere between five thousand and ten thousand—several times today's population even when you add in the white people. The long-ago population is estimated based on lidar aerial images that reveal thousands of collapsed stone habitations of all sizes.

Right among the pines in his backyard, Swetnam has three rubble mounds. Walking by them, you might not think twice. They just look like neglected construction debris—which they are, in a sense. Archaeologists call them collapsed fieldhouses—single-family stone huts that were likely occupied in summer, next to the crops. In contrast to the modern Jemez people, down in the hot lowlands, the ancestors preferred to live on the forested uplands, where they must have been better able to grow their corn, beans, and squash. Below Swetnam's house is a long stone-lined terrace, proving that crops were grown right here. In a bedrock slab next to one of Swetnam's fieldhouses they carved a round basin the size of a deep wash basin, much too deep for a *metate*. A constellation of small pits ("cupules") surrounds it, interconnected by faint traces of channels. Metates—hollows worn into rock over years of stone-grinding corn kernels—are pretty common in the region, but this hemispherical basin type is rare, and a mystery. Anthropologists attempting to interpret it can only offer vague guesses involving water ceremonies.

Over several centuries, five thousand people would need a lot of firewood for cooking and heat, and more wood for construction. To get it, they deforested their village sites for a few hundred yards in every direction. They also cleared patches of forest to make space for crops. It seems that this great patchwork, or network, of clearings prevented forest fires from spreading far.

Undoubtedly, the people also set forest fires, both on purpose and by accident. Burning was so universal among Indigenous people of the West that we have no reason to doubt that it happened here, serving here as elsewhere to thin forests, reduce understory fuel buildup, and push the fire regime in a low-severity direction. However, in New Mexico the practice largely ended long ago—by the year 1692, when the Spanish rounded up the entire (horrifyingly reduced) population into the lowlands near missions. By the time ethnographers came around to record informants telling them of the old ways, very little reached their ears about intentional burning. So firewood-gathering remains the best-known practice that served to manage forest fires in New Mexico.

It was powerfully effective. Intense fires since 1990 have overrun pueblo ruins, showing that vigas char, stones crack, and obsidian artifacts can either melt or pop like popcorn, turning into pumice—obsidian and pumice being alternate forms of rhyolite lava with different life histories. If any of these things had happened in the old days, evidence of it would surely have turned up; but it has not.

Swetnam and colleagues (including Ellis Margolis, whom we meet in the next chapter) have over time compiled a fire-scar record of the Jemez Mountains that's as thorough and complete as any in the West. They found four phases in the fire history here. Prior to 1680 (i.e., under Puebloan Indian management) fires were numerous, small, and predominantly low in severity. This pattern weakened after 1590, as the Spanish gradually drove the Indians out of the highlands. Then from 1680 to 1880, with less than 20 percent of the Puebloans surviving, and all in lowland towns, humans did not manage fire. Climate ruled, and lightning ignited. Fires were fewer and some of them were much larger, especially in dry summers following a year or two of above-average precipitation that grew more shrubs and herbs, creating more surface fuel. The years 1748 and 1806 burned the most area, all across New Mexico. In the Jemez Mountains, those two years saw fires much wider than Las Conchas—but the fires were still predominantly surface fires that most trees survived.

In 1879, the railroad arrived from the East. Suddenly, there was a way to ship sheep and cattle to market. Industrial-intensity grazing took off immediately, eliminating the grass layer. In many places several inches of soil eroded off the top, preventing small plants from returning even after the sheepherders abandoned the overgrazed lands. Without undergrowth beneath them, trees remained too far apart (thanks to lingering forest-thinning effects of frequent fire) to carry fire, so from 1880 to 1990 there were very few fires. The Forest Service began putting fires out, as well. The forests filled in. Tree crowns started to grow close enough together to catch each other on fire.

Then the climate heated up, and droughts struck. From 1990 on, new regime: big, bad fires.

Far to the northwest, the Rocky Mountain Trench is the longest nearly straight deep valley on the planet. Conspicuous from space, this striking feature stretches a thousand miles, the length of British Columbia and a short way into Montana, where it broadens and ends in Flathead Lake, the largest natural freshwater lake in the West south of Canada. Both flanks of the trench rise precipitously to peaks of 11,000 to almost 13,000 feet, from a valley floor at 2,500 feet. Though scoured out by Ice Age glaciers, the trench is where it is because of major geologic faults (like most long, straight valleys on earth). Its flat floor cradles rivers, of course, eight of them in series, with some flowing southeast and some northwest: pairs of them slip away from each other at invisibly low divides, or collide head-on, then shoot off eastward or westward through a side canyon. Those in the northern trench flow to the Arctic, the southern trench to the Pacific.

The southernmost trench floor, on both sides of the international border, is broad, fertile but fairly dry, and rich in vacationers. Here the Kootenai Salish grew tobacco for hundreds of years, and apparently set a lot of fires. Greg Greene found scars of fifty-three fires—a record for a single ponderosa pine. The surrounding mountain slopes are thickly forested except for recent clear-cuts and a few burns. They're deep green, dappled in fall with the golden yellow of western larches.

Greene is studying stands of Douglas-fir, western larch, and ponderosa pine in the southernmost trench. He had planned a straightforward fire history study, but as he examined the sequences of tree rings in his cross-sections, they posed a grim conundrum. Initially, they seemed illegible in tree-ring terms. The newest fifty years of rings didn't match the thick-and-thin sequences well known from elsewhere in the province. They didn't even match each other. Looking deeper toward the centers of the older Douglas-firs, he began to recognize sequences from before 1950, and his computer confirmed those matches—but there weren't a sufficient number of rings from there to the bark, not enough to count up to 2017. Five, ten, fifteen, as many as thirty rings that should have formed within the last fifty years were simply missing from many of these Douglas-firs. "A lot of trees have stopped growing entirely," he told me. A few died, but most were somehow hanging in there, not producing significant new wood tissue for years at a time, just a handful of new needles, waiting vainly for conditions to improve, not quite dying. Greene calls it "full-on stagnation—staying alive without growing."

In the arid Southwest, it's perfectly normal for trees like western junipers to have a lot of rings missing. Not so normal for Douglas-firs in the Northern Rockies. Greene has an idea of how the particular history in this area of British Columbia could be producing this result: it's a microcosm of a pattern we see all over the West.

These forests were high-grade logged between the 1890s and the 1930s. The loggers took the biggest trees, and they preferred pines, whose pitchy lower trunks resist rot when in use as railroad ties. Logging let sun stream into the stands, fostering a huge cohort of new seedlings, mainly Douglas-fir, the less drought-tolerant of the two species. The region's burgeoning population of settlers took up putting fires out, so the new cohort never got thinned by fire, and remains a stagnating dog-hair stand of predominantly small trees to this day. "There are hundreds of thousands of trees twelve feet tall that are 120 years old. There's too many trees, and you add climate change on top of that, you get entire stands that just stop being productive. They're just able to survive with the amount of rain that's coming, with increasing mortality over time. These are pretty young trees, a hundred years. They shouldn't be dying." Or perhaps a lot more of them should be dying, so that at least some of them would have a chance to thrive.

We shouldn't be surprised if that starts happening before long: severely weakened trees like these are often vulnerable to bark beetles and fire mortality.

The biggest trees, born before that first harvest, are growing especially poorly. In many situations, ecologists find that bigger plants have a great advantage over smaller ones because they have a longer reach with which to grab both sunlight and deep soil moisture. They call that "asymmetric competition." Here, Greene sees the opposite effect, "inverse asymmetric competition." It may be that the shallow but very numerous roots of the pole-sized trees are intercepting rainfall before it can soak to the depth where the older trees' roots are.

The stagnation has everything to do with water shortage, but may result more from the competitive structure than from climate change. Over the past century there's been a nice *upward* trend in the region's precipitation. But these stands are in a rain-shadowed valley with less than sixteen inches of annual precipitation, which isn't a lot, for a forest. And remember the two reasons why warmer temperatures make things effectively drier for a tree: stronger evaporation, and earlier snowmelt. An increasing portion of the winter precipitation is draining away from the site before the trees become photosynthetically active in the spring.

Similar stagnation in the late twentieth century is seen in ponderosa pines in Arizona and New Mexico. Lodgepole pine, and sometimes ponderosa, notoriously form ridiculously dense stagnant ("dog-hair") stands where growth rates grind to a near halt. In the Southwest, tree-ring studies find that competition from dog-hair infill may cause the old ponderosas that tower over them to also stagnate, in much the same pattern that Greene describes. It isn't as surprising in the Southwest, a very water-limited region that has been trending drier. Greene has searched the literature in vain for other tree-ring studies describing stagnating forests north of Colorado.

Two-Needle Piñon Pine

STRIEBY

5 THE BLEEDING EDGE

The reason people are interested in this part of the world is we're on the edge.

—CRAIG ALLEN

The highway east from Santa Fe, New Mexico, snakes up a narrow defile into low mountains until it reaches Glorieta Pass beneath the Permian sandstone cliffs of Glorieta Mesa. This is the path of the Super Chief, the legendary flagship passenger train of the Santa Fe Railway (and now Amtrak). Before that it was the path of the Santa Fe Trail. The trail brought supplies to the Spanish colony of Nuevo Mexico; then it brought Anglo settlers who began to move in on the Spanish; then it brought General Stephen Kearny and his Army of the West, who frightened the Mexican governor out of Santa Fe at the beginning of the Mexican-American War; then it brought Confederate and Union soldiers for the westernmost major battle of the Civil War. A pueblo here, called Pecos, may in its heyday have been the biggest population center of Puebloan peoples. Their downfall began with the very first European invasion of the region, under Coronado, who went through Glorieta Pass in 1540.

The Civil War battle of Glorieta Pass was a stalemate in terms of casualties, but it put an end to Confederate hopes of dominating the West when a detachment of New Mexico Volunteers slid and roped each other down the cliffs of Glorieta Mesa to surprise the Confederate rear guard in charge of supply wagons. The Union commander, Major John Chivington, ordered all the supply wagons to be burned and some six hundred Confederate mules bayoneted, to conserve ammunition. (Chivington later established his own special place in infamy by commanding a massacre of an Indian encampment of women and children.) The Confederate colonel, with his supplies and his steeds gone, decided the West was a lost cause.

Seventeen years later, in 1879, the railway line was stitched through Glorieta Pass to Santa Fe—the beginning of the end of natural frequent-fire regimes in New Mexico forests.

Glorieta Pass is not so much a gap in a great mountain range as it is the terminus of a great mountain range, the Rocky Mountains. Here the Proterozoic basement of the Sangre de Cristo Range gives way to broad sedimentary mesas of modest height. The Continental Divide is not here, nor is any major divide, as both sides of Glorieta Pass drain to the Gulf of Mexico. You can think of Glorieta Mesa as a ramp from the Great Plains up onto the feet of the Rockies. Grasses characteristic of the buffalo plains grow here, playing a key role in the fire ecology.

Tree-ring scientist Ellis Margolis made a breakthrough here, reading the annual rings of juniper trees. Previously junipers had often been dismissed as almost illegible and lacking much of a fire-scarred population. Margolis has driven me up onto Glorieta Mesa, with his black-Lab-mix dogs, Luca and Liz, panting in my ears from the back seat. The mesa is cattle range country, with expanses of grassland and swaths of piñon-juniper savanna and woodland punctuated with a few taller pines, ponderosas.

A slender, dark-bearded man, Margolis grew up in Denver, did graduate work with Tom Swetnam in Tucson, and spent time in a couple of flammable eastern pine ecosystems, in upstate New York and the Florida

Panhandle. On Glorieta Mesa he does both theory and practice: he studied and figured out the fire ecology of piñon-juniper savanna, and he also got to plan and oversee seventy thousand acres of forest restoration.

Margolis cracked the juniper code basically by working at it harder and longer. Most southwestern junipers are next to impossible to read because they grow so slowly and irregularly. The annual rings are minute and close together; many rings are missing (i.e., no growth that year) or they're missing on one side of the tree but present on another. Sometimes one side of the tree produces two rings in one year, when it has two rainy seasons separated by drought. So forget cores. A core doesn't give you enough sides of the tree; you need to study full cross-sections, following a ring all around the section to see what happens to it on the other side, until you know the ring patterns well enough to crossdate them. You need superfine sandpaper—fifteen-micron grit—to give you a high polish that makes the tiniest rings visible. Gradually you can figure out where there are missing rings and extra rings.

In some parts of the West it appeared that junipers reach old age only where they avoid fire—for example, by growing in crevices in rim rock, so that from one old juniper to the next there's nothing to carry fire, only bare rock. Junipers were germinating and growing on richer sites, but those sites carried fires to them that tended to kill them before they reached two hundred. Yes, junipers count old age in the thousands, and often still look like bushes as centenarians. Though a few fire-scarred junipers had been seen (even before Margolis), people still thought of junipers as trees unlikely to survive fires. After all, most of them bear dense foliage down to within a foot or two of the ground—a sure recipe for incineration, or so it would seem.

Nevertheless, piñon-juniper communities fall into three categories, and one of them is savanna, which means grassland with scattered trees. In most of the world's savannas, frequent grass fires maintain the grass by eliminating tree seedlings as they encroach, except for a lucky, scattered few trees that somehow make it to fireproof height. Margolis thought he might be able to show this was also true of P-J savanna, if only he could find enough fire scars. "Craig Allen, other people, they've been up here, they didn't publish anything off of this, because it was hard as heck to find fire scars." The places to look for fire scars, on Glorieta Mesa, turned out to be in the ravines that carry freshets during rainstorms. Wood debris tumbling in the freshets would pile up against the uphill sides of junipers and burn hot during the next fire, damaging the bark. Once scarred, the tree would scar further in the next fire, and in subsequent fires. He sawed scar sections out, and after hours and weeks and months of painstaking study confirmed that scarred junipers and pines across Glorieta Mesa share many of the same fire dates. That told him that the fires were extensive, and even the old trees without fire scars must have also survived those fires.

What makes it so different here? Why are fires so rarely able to scar trees? How are trees able to survive fires when they grow dense crowns just a few feet from the ground? "Because *this* is your fuel," says Margolis, batting at some wispy stalks of drying grass, "this super light grass that's not going to punch through the bark of, of *any*thing."

The grass of this savanna is blue grama, *Bouteloua gracilis*, mixed with lesser amounts of similar grasses. Blue grama and buffalo grass dominated the shortgrass prairie system that supported the bison, also known as American buffalo. That's the high, dry, western belt of the Great Plains, butting up against the Rockies on the west and merging eastward into the rainier, lusher tallgrass prairie. Tallgrass produced far more biomass, but the scruffy, durable shortgrasses were better able to nourish tens of millions of half-ton grazers, thanks to high protein content. Though capable of reaching twenty inches in height, blue grama was typically one to four inches tall where subjected to bison in combination with scant rainfall. It thrived under the abuse. The trampling herds repaid shortgrass with lots of urine and manure—good high-nitrogen fertilizer!—and the shortgrass recycled the nitrogen into protein, which nourished the bison, and so on.

Thanks to the fineness of the grasses and to the unrelenting winds across the mesa, the frequent lightning fires here moved fast and not very hot. By the time the piñons and junipers were several feet tall, they had dense crowns whose shade cleared grasses from the patch of ground under them. The flames might be a foot or two high in the grasses, but often right under the tree there was hardly any flame, not enough to torch even these very low crowns.

In consequence, piñon-juniper savanna was a frequent low-severity fire regime. You could say it's the same fire strategy as ponderosa pine, but achieved with the opposite tactics: ponderosa has high, sparse crowns and thick bark, and drops lots of flammable stuff on the ground to *foment* surface fires which ensure that ladder fuels won't grow up beneath them. Juniper and piñon pine have low, dense crowns, using shade to *weaken* surface fires directly under them. Either way, as long as the fires are frequent they'll rarely be hot enough to overcome the dominant trees' fire protection.

The savanna tactic depends on there being plenty of insubstantial grasses like blue grama. They seem to require the late-summer rains, called monsoons, that move north from Mexico and hit P-J country mainly in New Mexico and adjacent parts of Colorado, Arizona, and Texas. The monsoon index—the ratio of July-August-September precipitation to total annual precipitation—forms a gradient across a map of the West, from around 50 percent in New Mexico to negligible along the Pacific.

And then we saw junipers appearing to sabotage ponderosa pine's strategy. Dense groves of young junipers were clumped around the feet of the ponderosas. "Birds sit on the trees," Margolis explained. "They crap the juniper berries they've been eating, the seeds germinate. So now, when there's a burn, you'll blow the whole thing." No, I'm not seriously calling that an adaptation that arose because it benefits juniper by getting rid of the taller trees. The baby juniper groves grew under the pines because the railroad brought the sheep that ate the blue grama, interrupting the steady series of fires that the trees' fire strategy banks on. "After the railroad there were just too many creatures up there and it couldn't burn any more."

Remember, though, that savanna is just one of three piñon-juniper community types. The other two are much more widespread, and have completely different fire regimes. The archetypal P-J community is the piñon-juniper woodland. Piñon-juniper shrublands, on the other hand, include areas many ranchers like to think of as their cattle range—sagebrush steppe-grassland, until junipers move in. Speaking objectively, junipers and/or piñon pines did spread across a lot of sagebrush steppe over the last century, but ecologists see evidence over many centuries of multiple sweeping invasions, and equally sweeping diebacks.

Piñon-juniper isn't a marriage of just two species. There are four species of piñon pine and at least six species of juniper that hyphenate with each other. The piñon pines are all closely related, but they divide the Southwest up into territories, so a piñon-juniper community typically has just one piñon species and one or two juniper species. On Glorieta Mesa (as in most of northern New Mexico) it's two-needle piñon with Rocky Mountain juniper and one-seed juniper. Rocky Mountain junipers provide Margolis's cross-sections, since they grow much bigger and older than one-seed juniper. Piñons and junipers aren't closely related to each other, of course, but in their hyphenated communities they begin to look kind of similar, with gnarly trunks and clumpy crowns of dense, dark green foliage. Close up, their foliage has little in common—needles on the pines, and sprays of tiny scalelike leaves on the junipers. Though both are conifers, junipers don't bear what anyone would normally call a cone. They bear their seeds in dry, blackish to bluish berries, which are a major food for some birds, especially cedar waxwings.

Piñon pines are among the pines whose seeds clearly evolved to attract and to reward eaters. Piñon nuts are a major food for native peoples in the Southwest. Though smaller than Italian pignoli, they are just as tasty. Various birds and squirrels eat them and also disseminate them. A crestless bright-blue jay, the pinyon jay, specializes in them in a mutualism similar to that of Clark's nutcrackers with whitebark pines (chapter 8), with the difference that piñon pines do not depend on jays alone. (Pinyon jay is the Ornithologist Union–approved spelling for the bird. Forest scientists, on the other hand, tend to stick with the original Spanish spelling, piñon, as well as pronunciation, peen-YAWN. At least the New Mexico scientists do. The Spanish legacy is powerful in New Mexico. Margolis speaks Spanish at home with his wife from Mexico City and their son.) Pinyon jays live in large flocks, and cache piñon seeds en masse—planting piñon pines with a level of efficiency humans could only dream of. Their populations cycle up and down in response to piñon pine mast years (years of strong cone production), which come along roughly three times a decade. In the off years, very few baby jays make it to adulthood. The sight of plentiful green piñon cones stimulates the gonads in both sexes of pinyon jays, gearing up for a good crop of babies. However, mast years are somewhat tied to precipitation, and have been fewer in the last twenty years.

Overall, pinyon jay populations have declined by about 85 percent over the past fifty years. That can't be blamed on the total number of piñon pines, which was increasing up until 2002, so researchers are tracing the jay's decline to subtler long-term declines in the *quality* of piñon habitat. I saw no pinyon jays while I was in the Southwest.

Piñon-juniper combinations characterize the borderlands between forest and nonforest across a vast Southwest realm east of the Sierra Nevada and south from Idaho. Junipers are the most drought-tolerant conifers in the West. Piñons are a little less drought-tolerant, but still a little more so than ponderosa pine. In other chapters, I describe ponderosa pine (or limber pine in a few locales) as the tree of the lower timberline, the tree on the climatic edge between where forests can and cannot grow. But that isn't true everywhere: in the northern half of the West, junipers sometimes line the ponderosa pine edges at lower timberline. They tend to grow as shrubs there, which perhaps gives us an excuse to still think of it as a ponderosa pine *tree*line. In much of the southern half of the West, on the other hand, the lower timberline trees are piñon pines and junipers. Piñon-juniper woodlands form an entire belt that typically intervenes between treeless country and ponderosa pines. In a warming climate, if ponderosa pines are the edge, then piñon-juniper is the bleeding edge.

The edge hemorrhaged explosively in 2002, in the extreme third year of severe drought across the Southwest. There had been years equally dry earlier in the twentieth century; these three years were dry again, and now a few degrees warmer, increasing the stress on trees and provoking bark beetles to an extent that had not been seen in the written-history era in the Southwest. Craig Allen speaks of "hotter droughts" to make the point that this is a clear effect of climate change.

The main deadly bark beetle on piñon pines has funny names: the piñon ips, or in Latin, *Ips confusus*. Like our other bark beetles, it is always around in small numbers, normally doing little damage, but then in response to drought it erupts. As with other bark beetles on pines, you can recognize its attacks by the little lumps of pitch that ooze out through the holes where adult beetles have entered. The 2002–2003 outbreak killed nearly all of the piñon pines in some tracts of piñon-juniper woodland that range up to county size, turning them into just plain juniper woodland. The area from Santa Fe to the Four Corners was hit especially hard, while Glorieta Mesa—and in fact most of the vast piñon-juniper realm—went almost unscathed.

Margolis remains fatalistic about piñon's chances, and even more so about ponderosa. "Physiologically, junipers are going to be the one that I put money on. We haven't had *Ips* up here; eventually it's going to hit the piñon up here." When given the responsibility of planning a restoration treatment for seventy thousand acres, he decided to reduce density without trying to favor pines. "My personal goal is, bring the process [fire] back and we'll see where it goes. We're not going back to 1880. Most of the ponderosas on this landscape are on death row. We're at the lower elevation limit of ponderosa. Piñon and juniper, that's what you see regenerating. These are going to be P-J landscapes."

He wants to bring fire back before the wrong kind of fire comes back uninvited. "We don't want a fire to wipe this whole thing clean. It's so dense now, and the canopies are so connected, that if you had a fire with a south-west wind crankin', you'd lose everything. We want to break up the present fuel continuity."

The toughest question for Margolis was, "How am I going to get grass to grow here?" The mesa is still a grazing allotment—for cattle now, not sheep—so in addition to all the usual constraints on when prescribed burning is permissible, the window of opportunity also requires the cattle to have been rotated elsewhere months beforehand, allowing grass to grow back, to provide fuel. Worse, much of the terrain simply can't grow grass anymore because the sheep overgrazing a hundred years ago led to erosion removing the topsoil. Shrubs, junipers, and piñons can get established in the arid raw dirt that remains, but grass and flowers cannot. Erosion "pushed parts of Glorieta Mesa into another realm, a place where they will now be P-J woodland [in place of P-J savanna] because there's no soil anymore and no grass . . . You can't restore stuff like that . . .

"As a dendrochronologist I get to look back in time. I'm still surprised at how much these landscapes have changed since, you know, circa 1900."

It's impossible to predict precisely which of today's forest sites will be able to support a forest a hundred or two hundred years from now. Nate McDowell is one scientist who has developed computer models and published predictions, and they are very grim. He says his several mutually independent models indicate "that conifers will be gone from the southwest by 2050." When pressed, he concedes that "we should expect some refugia, some legacies that enable some survival."

Does that mean there's no point in forest restoration in New Mexico? No. Restoration is well accepted in the Santa Fe area because people want to reduce the chance of severe wildfires that cause flooding and damage water systems. If restoration forestry can accomplish that for thirty years, and postpone deforestation for the same period, that's most of a human lifetime. It will have been worth doing. But ultimately, restoration efforts in the land of piñon-juniper rest on a hope that McDowell's models are exaggerating, together with a hope-against-hope that before the twenty-second century arrives humanity will have halted if not reversed the curve of climate warming.

To see a fine well-preserved example of the piñon-juniper woodland type, look no further than Mesa Verde National Park, Colorado. But maybe you'd better look back, back into the mists of time, to the twentieth century. In a fifteen-year span from 1989 to 2003, half of the park's woodland went up in wildfires. Then between 2002 and 2005, bark beetles killed half of the remaining big, old piñon pines. This cannot be a typical rate of mortality for these trees. The older trees in those woodlands were four hundred, even five hundred years old, and no one could even find fire scars on them. None. From that we can conclude that fires in those woodlands are normally of stand-replacing severity, but were very rare—so rare that most of the park's piñon-juniper woodlands hadn't burned in 200 to 550 years—until recently.

When I meet park ecologist Tova Spector at the park headquarters, we don't have to take a single step to view fire damage; we just look across a narrow canyon at Spruce Tree House, one of the park's signature Puebloan cliff villages. Spruce Tree House is missing its hat—the old-growth piñon-juniper woodland that until 2002 covered the mesa top directly above it. A five-minute walk gets us there. Where this fire was an imminent threat to the headquarters buildings, no effort was spared in fighting it. Red traces of fire retardant stain trees both in the incinerated patch and in the saved patch next to it.

Looking at the two patches up close, I see the bad news. Seedlings and small saplings of piñon and juniper are plentiful in the intact patch, and nonexistent in the burn. Weeds, on the other hand, are negligible in the intact patch, but in the burn they are more than half of what has come up since the fire—non-native invasive herbs and grasses, in botany speak.

A Californian by birth, Spector got much of her professional experience in Florida. Her Park Ranger cap can barely constrain her thick gray curls or shade her bright blue eyes. "Best guess," she tells me, "we think it takes over a hundred years to get trees back into these burns. First the shrubs move back in. That takes about fifty years, and once shrubs are established, they serve as nurse plants," shading patches of ground so that the soil stays cool enough and moist enough through summer for a tree seedling to survive. As far as we can tell, reestablishing piñon-juniper woodland after it burns down was always a very slow process. Some former woodland in the park documented to have burned in around 1880 is still brush, with a few scattered trees.

A couple of things make it dicier now: it's hotter today, and we've got these weeds to contend with. Not only do the weeds compete with tree seedlings, but the worst one, cheatgrass, is notorious for spreading fires. As long as the burns are full of cheatgrass, there's little hope that nature can complete the one hundred fire-free years that may be needed to bring back piñon trees. The park gets several lightning-sparked "single-tree fires" every year. If a lightning-lit tree, or big shrub, or even a dead snag is surrounded by cheatgrass, the fire is less likely to settle for just a single tree.

Cheatgrass has been invading the United States since the 1890s. It has replaced native sagebrush communities in some places, and infests more than 60 percent of those that remain. Land managers haven't had a truly effective tool in their toolbox to fight it with, but when I spoke with Spector she was eagerly awaiting the arrival of a new tool that looks like the most promising to date.

It's a bacterium, a strain of *Pseudomonas fluorescens*, a species ubiquitous in soil and also on people, animals, and almost everywhere. Even the particular strains that suppress cheatgrass roots seem widespread, albeit sparse, in western soils. A bacterial treatment thus wouldn't add an organism that isn't already there; it would just augment the quantity of that strain. To be effective, the bacteria have to be carried by rain down into the soil and have to survive and reproduce there, but their numbers actually dwindle even while their effect on cheatgrass increases for two to four years. They fade out completely by year six, when they can be reapplied if necessary. One of cheatgrass's biggest competitive advantages is that its roots grow through the winter; with its root growth suppressed, it can't compete.

Spector and other land managers have their fingers crossed that this therapy will be a godsend for their cheatgrass problems. At best, though, it will be used in national parks and other high-value sites, and possibly in landscape-scale stripes—a new kind of fuel break to limit fires in compromised sagebrush ecosystems. No one is dangling any promises that the bacterial culture will spread on its own to reduce cheatgrass West-wide, and spraying it West-wide would be way too expensive.

U.S. Department of Agriculture soil scientist Ann Kennedy worked toward this day—the day *P. fluorescens* biovar B strain ACK55 drops—for thirty years. Testing . . . testing . . . testing . . .

She had noticed a patch of soil where grasses were doing poorly, so she cultured bacteria from that soil, yielding 7,000 distinct strains of *P. fluorescens* and its close relatives. The first screening procedure found 3,150 of those strains that suppressed cheatgrass roots in the lab. (As well as roots of two other much-hated invasive grasses.) Nineteen hundred of those flunked the next test by also suppressing either agricultural wheat or a beloved native bunchgrass. She tested the remaining 1,250 on more than a hundred other kinds of plants, narrowing the field to 625 strains that did no harm. The next round of tests was similar, but slower, conducted in pots of soil rather than petri dishes of agar. Just sixteen candidates remained after that round, and eleven of the sixteen failed the next round by inhibiting members of the fungal kingdom. Then it was on to the animal kingdom for the five remaining strains. They held up under the stress, doing no harm to ladybugs reared on macerated grapes, nor to daphnia (water fleas), nor to rats, nor did they "compromise the integrity of lettuce or wheat protoplasts." Tested for survival in "ditch, lake, ocean, and river waters," one scored about as poorly as the next—and poorly was the correct answer to that test question. Five candidates were at last ready for testing out in the field.

I'll skip the rest of the details, but I can assure you that cheatgrass is such a disaster that no one involved with it has the slightest doubt that it was worth all of Ann Kennedy's trouble. Cheatgrass has shown again and again that it can completely replace native plant communities via a frequent-fire feedback loop that precludes the native plants' reproductive strategies and perpetuates more cheatgrass. The main victim, so far, has been sagebrush in the Great Basin, where fires are already far larger and more frequent than they were without invasive grasses. But Mesa Verde is not the only setting where cheatgrass looks like it could possibly replace forest or woodland.

Some scientists warn of similar effects in California dry forests where it is well established.

Forest ecologists often say that drought kills trees, and that bark beetles are often the "proximate agent"—the last thing pushing a tree over the edge. Telling whether it's pests or solely drought that's killing trees is harder than you might think. In California in 2015, the third year of extreme drought, I heard that while beetles were killing pines and firs, incense cedars were succumbing to drought plain and simple. In 2018, a Nate Stephenson crew studied the issue more closely and found that cedar bark beetles are a very significant mortality factor, and that unlike the pine beetles, they prefer to attack small trees.

For a tree, no matter how dry last year's drought was, this year's drought can always be worse, simply by being hotter. The force that sucks water out of trees into the air, called vapor pressure deficit, is a simple formula of air temperature and relative humidity. As temperature increases, vapor pressure deficit increases exponentially. "If you take two sets of trees, experimentally," Craig Allen explains, "and you turn the water off to both sets at the same time, but one set you warm, the warmer ones die faster. Globally, what we're seeing is that warmer droughts are killing more trees than droughts used to do."

Allen and his colleagues work hard figuring out the roles of two physiological mechanisms in direct drought mortality. Every tree is a bundle of countless tiny vessels, each carrying water up from a root to a pore on a leaf, where water evaporates into the air. This evaporation isn't just incidental; it's a necessary part of photosynthesis, the chemical reaction that's the basis for life on earth. Water is one of the two materials for photosynthesis. The force that draws water up into the leaves for photosynthesis is suction, starting with evaporation from the leaf pores, each of which is at the top of one of these vessels running all the way to the roots. "Where water transpires out, it pulls one molecule out of the leaf surface, which pulls the next one in line, all the way down to the rootlets, that have to pull it away from the soil."

This suction gets staggeringly powerful in the case of a 380-foot redwood or Douglas-fir. But it has limits. If the vapor pressure deficit at the top rises too high for the supply of water from the bottom, the tiny vessel breaks. "Embolism" and "cavitation" are two words for this. Once it breaks, that particular vessel may never function again. "If you lose about fifty percent of these conduits, you're gonna die."

Trees try not to let that happen. To prevent it they shut their leaf pores. Where a leaf pore is shut, no photosynthesis can take place; that part of the tree is dormant, it isn't sustaining life. Most conifers in the West spend a good part—maybe I should call it the bad part, but at any rate a substantial part—of every summer largely dormant. They wake up again when rain falls, or even when fog wets their leaves. In Mediterranean climates near the coast, summers can be consistently dry but plants can get a lot of their photosynthesis done in spring and fall, or even winter. A select few live where their roots reach the water table, typically in a valley with a stream or at least a seasonal stream. But when drought forces trees into dormancy for most of what should be their growing season, they starve. We call it that because carbohydrates from photosynthesis are the "food" they use to build themselves, and the supply of food has been cut off. Thus, when a tree dies of drought without a pest or disease agent, it's thought to be a case of either cavitation or carbon starvation. Either can happen; but a tree approaching death is weakened, and typically falls prey to pests or diseases first.

Allen looks at the drought effect of rising temperatures as a global phenomenon. "On hot days everywhere, from the tropics to the boreal forest, trees close their pores. But in places like this, where you get protracted droughts that last weeks, months, years, chronically, trees are photosynthesizing less, growing more poorly, and investing less in plant defenses.

"Ironically, in all those places like central Europe and eastern North America, people aren't too worried yet about drought- and heat-induced tree mortality . . . Here it takes a year or two of stress to kill trees because, well, year-long droughts, these trees have seen it before, many times. In the East, if you go two months without precip, stuff is going to be gasping."

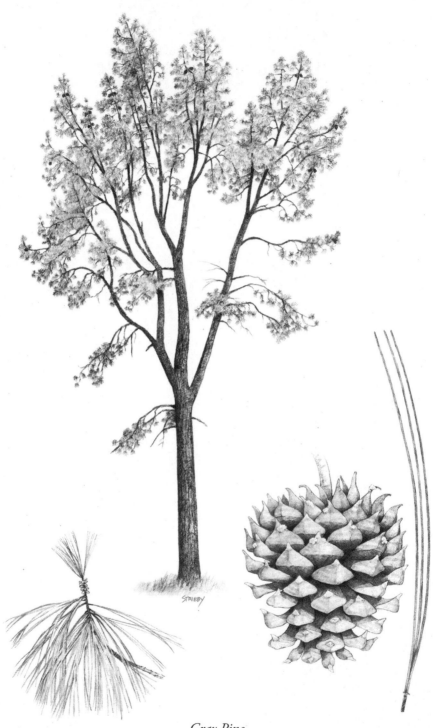

Gray Pine

6   THIN AND BURN

On the ashy ground, an inch-diameter hemisphere of hot pink catches my eye. It's a plastic ball like a Ping-Pong ball with a pink hemisphere, a white hemisphere, and a little smoke-tinged hole melted in one side. "Oh yeah, we ping-ponged it here." Ellis Margolis explains to me that prescribed burning as it's usually illustrated—workers in hardhats and yellow Nomex suits carrying drip torches dripping flames through a smoldering forest understory—is hopelessly inefficient when you have thousands of acres to burn. Instead, you drop these little bargain-basement incendiary bombs from a helicopter. Ping-Pong balls full of potassium permanganate ride in a chopper's hopper that drops them one by one, injecting each one on its way out with a bit of antifreeze (propylene glycol). After half a minute, having reached the ground, this mixture spontaneously combusts, to burn for another half minute or so. In other forests, ping-ponging is sometimes the method of choice where rugged terrain makes hands-on torching risky for personnel; sometimes a still coarser tool is used, the helitorch, where the helicopter carries an oversize drip torch.

We are walking through one of Margolis's prescribed burns. It's one that got away from them, in a sense. It killed most of the trees within one patch a couple of acres in size. It doesn't bother him; it's a negligibly small portion of the entire plan, and it wasn't an especially fine or old stand. "Prescribed fire," he says, "is a blunt tool." Nearly every scientist I've talked to has offered me some version of that warning when the subject of Rx fire came around. And they generally shrug off concerns over particular Rx fires that burned a little hotter and killed a few more trees than intended. "Fire creates heterogeneity—it always did—and that's a good thing. We've dug ourselves a big hole, and it's hard to get out of it with precision."

On the other hand, I've heard plenty of horror expressed by members of the public about dead trees in prescribed burns. I think people are easily sold on the idea of reducing wildfire risk by clearing out the brush and saplings, and they may miss the part about reducing the density of *trees*. Some trees are supposed to die; that's the plan. The density of frequent-fire forest stands prior to white settlement was often really low—not just a little bit lower than today, a lot lower. Twenty to sixty trees per acre, commonly. At twenty per acre you might see it as scattered pines rather than a forest. Pioneers saw it as "pine openings" and the first surveyors in Oregon sometimes called it meadow. Plenty of photos from the 1930s and earlier show unevenly scattered pines in places that subsequently turned into continuous forest.

Prescribed fires that under-kill are probably as common as ones that over-kill. Under-kill may fail to accomplish the goal—improving fire resilience—and it wastes time and resources. Prescribed fire is a blunt tool.

"Thin and burn," broadly speaking, are the two available "treatments" for reducing forest density. The burning part can be either prescribed fires (set and controlled by burn crews; also known as controlled burns) or managed wildfires (set by lightning and then allowed to burn for ecological benefit, subject to close monitoring and limits).

Thinning is a selective form of logging. When it's "forest restoration," it aims to leave the stand with a mix of tree sizes and spacing, while preserving the bigger trees of fire-resistant species. It's often ideal to thin first, so that the ladder fuels are removed and the fire risk to the more desired trees is reduced, and then to burn within a few years. Both thinning and burning have virtues that the other can't duplicate. Thinning alone increases the fine fuels on the forest floor, unless you go to the trouble and expense of picking them up and hauling them off. Burning alone may fail to reduce the forest density, especially where you need to kill some sizable trees, not just small ones. And it doesn't give managers as much control over which trees remain, or their spacing. (Heterogeneous spacing is often an objective.)

But in many particular locales, only thinning or only burning is feasible. Of the innumerable studies evaluating their effects, many found the desired effects accruing to burning alone, others that they accrue only to the combination, and some that they can also accrue to thinning alone. A few found no significant benefits at all, or even undesired effects overall.

Where would thinning with no burning be called for? "Close to development" is the example I hear most often. Obviously it's what most homeowners would be comfortable with on their own property, and probably anywhere within a few miles. It's amazing, though, how much that can change with local experience. They burn five minutes from downtown Santa Fe, New Mexico. They burn two blocks away from homes in Camp Sherman, Oregon. And in Florida and Georgia. Small, well-controlled burns, of course. Ideally, prescribed burning is desirable close to development as well as far from it.

Lisa Floyd-Hanna applies the "close to development" test to thinning piñon-juniper woodland (her area of expertise), and she also rules out burning. Piñon-juniper can be Rx-burned where it takes the form of a savanna—sparse trees with a carpet of fine grasses between them—but where it's a woodland it usually won't carry a fire. When it does—in high winds after extended drought—it will carry a crown fire that kills most of the trees. These woodlands persevered and thrived for millennia because their fires, though typically stand-replacing, were very rare, hundreds of years apart on any given site. The extensive fires at Mesa Verde are part of an ominous region-wide trend that pencils out to unsustainable high fire frequency in recent decades. Though thinning is the only known way to mitigate fire hazard in dense piñon-juniper woodland, and should be done close to valued resources, there's nothing natural about it and it brings several possible risks, such as the newly exposed piñons blowing down.

In the timber industry (no surprise here), thinning is quite popular—in a position paper from Sierra Pacific Industries, for example—and prescribed fires are not. (If it's a dry forest type, they always have to get rid of the flammable debris from thinning somehow, often by piling and burning to minimize costs and to avoid prescribed fires.)

Situations calling for burning without thinning, in contrast, are widespread. Start with all the places dense enough to call for treatment, but unavailable for thinning—wilderness and inventoried roadless areas; municipal watersheds; buffer zones along creeks; areas too far from roads; steep slopes, where mechanized logging leads to erosion; known habitat of sensitive species. Malcolm North, a Forest Service research ecologist, added up all the parts of California's Sierra Nevada where one or the other of those constraints apply; he added in areas lacking marketable timber, which is another frequent constraint. He concluded that only 20 percent of the watersheds his study was divided into "had enough unconstrained acreage to effectively contain or suppress wildfire with mechanical treatment alone."

Several studies calculate that treating 15 to 30 percent of the forest acreage is enough to help limit wildfires, provided that the treatments can be spatially arranged well; however, the constraints often hamstring your ability to arrange them. If you're limited to the feasible rather than the optimal spatial arrangement, the portion of total area you need to treat in order to limit fires substantially is probably at least one-third, or 3 to 4 percent each year.

East of California, much less of the burnable forest area is constrained. On the other hand, the portion where treatments can pay for themselves through timber sales is even smaller. Analyses have been done showing that treatments there would pay for themselves through the avoided costs of fires. That argument won approval from the cities of Santa Fe and Flagstaff, partly because they realized their water supply is inextricably tied to their fire issue.

North drew two conclusions about how to deal with the constraints. First, some of the administrative rules, like avoiding steep slopes, wildlife habitat, and wide buffers along streamsides, should be relaxed slightly. After all, those rules were written to protect ecological values back in a time (not so long ago) when the ecological harm of forest density wasn't well understood. Some of the rules were always meant to be temporary. Balancing those rules better against the overall integrity of the forest ecosystem would yield net ecological benefits. Of course, we should also expand treatments by subsidizing them where there aren't enough marketable thinnings to pay for them.

His second conclusion: a lot of the density reduction will have to be accomplished by fire alone. In some of the places where you can't thin, there's increasing interest in intentionally shooting for mixed-severity fire. Though a ponderosa pine forest is ideally maintained by low-severity fires, after eighty years of densification it may need something more intense to convert it back again. Natural presettlement forests were sparse, with a modest number of large, old fire-tolerant trees. Today, that modest number of large fire-tolerant trees has been reduced by logging and disease, while the space in between has filled up with lots and lots of saplings and trees of all sizes. You might be able to accomplish a low-severity prescribed burn by setting it at a damp enough time of year, but that would change the forest so little that it would be a waste of time and money. To restore conditions to anything like the original (in a location where thinning isn't in the cards), you need the fire to kill quite a few of those medium-to-large trees. That makes it a medium-severity fire.

Prescribed fire has its own set of constraints: costs, planning, paperwork, air quality, some of the same water quality issues as with thinning, liability when it's near development, the availability of skilled crews and helicopters ready to go when the window opens—that one, above all: the window. The right moment. Burns get all planned and approved, and then they wait and wait, sometimes years, for the perfect calm. Woody debris on the ground has to dry out enough to burn, but not so much that it will explode. (Fire managers go out and measure it, day after day.) Not too much wind, but enough wind to move stagnant air and smoke, especially from nearby populated valleys. A cool day. Rain in the forecast is a plus.

It's nine degrees as I venture from my sleeping bag into the brilliant blue October dawn at Sycan Marsh Nature Preserve, in south-central Oregon. My travels to learn about trees in trouble ended near where they began, in this almost nameless range of modest mountains forming the headwaters of the Klamath River. On a map of the range of ponderosa pine–dominated forests, it's one of the biggest blocks.

Cold in itself can't be a bad thing for a prescribed burn, but dry air tends to accompany it. As the day warms up, the relative humidity (water vapor content relative to temperature) inevitably plummets, according to simple arithmetic. Today's forecast humidity drops to an afternoon low of 17 percent. That would brush up against the limit of the prescription for the burn. The summer got near-record low rainfall, and the fall rains so far have amounted to only six tenths of an inch.

The Nature Conservancy owns this preserve and is in charge of this burn. The actual Sycan—a Klamath Indian term meaning "flat grassland"— is ten miles across and a mile high. On their part of it, TNC is conducting an experiment in preserving habitat simultaneously for precarious wildlife and for the precarious ranching way of life. The bull trout, tundra swans, and yellow rails cohabit with four thousand head of black Angus, as well as elk and deer. On slopes surrounding the Sycan, the preserve includes 6,453 acres of forest, largely ponderosa pine. That land is used to promote resilient landscapes using prescribed burns for training, refinement of techniques, and cutting-edge research. On hand today are thirteen scientists from three institutions, plus eight fire engines (the small kind, for forest fires) and several firing crews totaling thirty-four firefighters.

In command of all this is the burn boss, Katie Sauerbrey, a wiry young woman with a pierced nose and piercing blue eyes in a small triangular face accented by asymmetrically upswept hair. Her mission "is to build a fire army of ecological burners, because it's the right thing to do for the ecosystem. Fire is medicine for the earth."

The job of burn boss stretches from Incident Command System to choreography. (ICS sounds paramilitary, and is used by the military, but was originally developed for forest firefighting.) Crew members with drip torches to ignite the ground at their feet will enter stage right, staggered, walking in just the right direction so that the fire will move in front of them and never trap anyone from behind. The boss plans out those movements based on calculations from wind speed and direction, topographic slope, and the nature and moisture level of the fuels. She spells it all out in the morning briefing. "If you encounter the public, contact me by radio and I will interface with them." Her hard-ass side is on display, but I suspect that's an obligatory persona that will fall from her shoulders along with the tensions as soon as her burn-boss season ends and her job title reverts to manager of the preserve. It's a remote, quiet place; access in winter takes an hour at full throttle on a snowmobile. Sauerbrey will move to a small town for part of winter, while a young couple newly arrived here will try their luck as winter caretakers.

Crews for spring and fall prescribed fires typically spend their summers fighting wildfires; these opposite-seeming pursuits share a lot of techniques, since lighting backfires with drip torches is a common method for creating firebreaks around wildfires. A firefighter swinging a drip torch looks a bit like a gardener swinging a watering can, except that the fluid being sprinkled is flame. The torch is a cylindrical canister with a handle on the side and a long tube sticking out at the top. The tube has a wick at the tip to sustain flame, and a curlicue in the middle to block flame from spreading into the can. The fuel is a three-to-one mix of diesel fuel and gasoline.

One of today's crews sets itself off from most fire crews in that its members all have degrees in botany or ecology. After dinners here, bottles and jugs of bourbon and rye come out; small tumblers are filled and emptied, straight, no rocks. That's a diametric difference from fire (fighting) camps, which are strictly dry. Russ Parsons, a researcher from the USFS Fire Lab in Missoula, tries to explain to me how strict abstinence while fighting wildfires and liberal indulgence the night before lighting controlled burns are each perfectly appropriate to those respective activities, but his logic remains a bit fuzzy to me. He's six two, with a big grin, a bigger laugh, and huge gestures, the class entertainer, frequently lapsing into Scottish brogue just for kicks, or a fleeting attempt at eastern Kentucky. He and Adam Watts, who heads a team from the Desert Research Institute in Reno, both served in the Peace Corps; to get to know each other they trade shots of rye and bloodcurdling tales of their experiences in Ecuador and Guinea.

Two of the research teams are flying drones. The Nevada team has a six-rotor, six-motor monster by DJI, like a teenager's Phantom drone on steroids. Fire ecologist Kellen Nelson designed and built its twelve-pound instrument payload, including two shiny one-liter vacuum spheres that will open on command and suck in smoke from the heart of the fire's smoke plume. The team will analyze its chemistry, and the payload will have also recorded its temperature and other details. The drone from the University of Montana will fly higher, taking high-resolution pictures that record the visible spectrum plus infrared. The Fire Lab team laid out grids for the aerial camera to zero in on, plus a dozen "fire behavior packages" and GoPro cameras all mounted on little tripods and enclosed in insulated, foil-wrapped boxes.

After the morning briefing we all head out to the edge of Brattain Ridge Burn Unit 2. This is a change of plans, switching the respective days for burning Unit 1 and Unit 2, so the Fire Lab crew runs around moving all their foil-boxed packages between units. When we're done, we reassemble, only to learn that Sauerbrey has canceled the burn for today. The relative humidity was running eight percentage points lower than its forecast. Sure enough, by midafternoon it hits 5 percent, which would be crazy low for burning. Forecast for tomorrow is similar but a few points more humid. We'll try again, switching to Brattain Ridge Unit 1, which has a little less fuel, making it safer to burn in marginal weather.

Next morning, the Forest Service asks us to hold off ignition until 11:00, for air quality reasons. Meanwhile, a wildfire has broken out fifty miles to the south, and all the "resources" (fire crews) in two counties have been called to fight it; that means that if our burn were somehow to escape its intended boundaries, no backup resources would be available. The humidity is a few percent higher than yesterday, good, but the wind speed is also a few miles per hour higher, bad, so that's a wash. At 12:15 p.m., the burn boss cancels for a second day. "You want to get a reputation for pushing the boundaries. You don't want to get a reputation for being reckless." Thirteen researchers, thirty-two fire crew members, seven fire engines, and one writer are sent to pack up their tents for now and come back, ever hopeful, five days later when an actual change in the weather is forecast.

With their hexacopter drone and its vacuum cylinders, the team from Reno has developed a superior technology for analyzing the chemistry of forest fire smoke. This research will become really interesting—and vitally important—once the results are in from a wide variety of vegetation types burning with a range of intensities, including wildfires. That's some time off in the future, but already they have started finding chemicals that no one had ever dreamed would go up in smoke. The technology should allow comparisons of the likely health effects of prescribed fires and high-severity wildfires.

In 2013 we got a striking first answer to that question. A NASA plane full of all the latest instruments flew around right in the plumes over three wildfires, including the Rim megafire, measuring the many kinds of bad stuff in wildfire smoke. That had never been done before because, well, because it's a damned scary thing to do. The actual wildfire smoke turned out to be three times worse than previous estimates, especially in the category of sub-micron-sized particulates, the size most hazardous to human health. Looking on the bright side, this tells us that wildfire smoke is likely to be many times worse for us than prescribed fire smoke—even for a given quantity of biomass combusted. I look forward to seeing more confirmation of this result; for now, we have a strong basis for gritting our teeth and saying "Yes!" to prescribed fire smoke. Some smoke now will mean less of that bad smoke in the future. It stands to reason: wildfires burn much hotter, and form much stronger convection plumes. Of course, when wildfires burn a lot of houses, the smoke is more toxic still, increasing the advantage of prescribed fire smoke.

As for the teams from Montana, the objective of their research is to develop detailed fuel maps and computer modeling tools that fire bosses in the near future can use to make good decisions and safe choreography. Fire models are used today to predict trajectories of wildfires but not much for prescribed fires because they lack sufficient detail. Fast, accurate fire models would go a long way to sharpen that notoriously blunt tool, prescribed fire.

A team from Reed College comes here with a mobile lab they call BioBasecamp. It's an adorable Airstream trailer covered with enough so- lar panels and lithium batteries to power all the instruments inside, which directly measure the degree of drought stress in living wood. For example, they measure fire damage to the xylem's ability to conduct water. This mea- surement could also go into the models. The team leader, Aaron Ramirez, tells me they're testing a hypothesis that trees are less likely to be killed by a fall burn than a spring burn if the late summer and fall have left the trees with their stomata closed, potentially averting cavitation damage.

Elsewhere, studies find that as trees get more drought-stressed over sev- eral years' time, they become more likely to die in a fire of a given heat and intensity. That suggests that thinning can help reduce fire mortality even when it doesn't prevent the fire; and that by burning after extended drought, a burn boss could achieve greater density reduction without needing a hotter fire, one more likely to escape. This mortality would be selective, culling the more troubled trees.

The models have to be able to adjust to a wild diversity of local vegeta- tion and climates. Every part of the West, every type of vegetation, warrants its own prescription. In Montana they say it has to be in spring, before the grasses and perennial herbs come up; they're so moist they make burning impossible. On Glorieta Mesa, New Mexico, on the other hand, grasses make burning possible; Margolis had to burn after the grasses came up. A hundred and ten years ago, when our society chose the wrong fork in the road regarding fire, there were grizzled fire-setters with local wisdom passed along from generation to generation. There are still plenty in the Southeast, not so many in the West, though some of the tribes carry on their traditional expertise in burning programs today. We're relearning the old ways, and filling gaps with modern technology as fast as we can.

That said, there are highly skilled and effective prescribed fire managers in the West by now. We just need a lot more of them on the payroll. In the United States, the area burned with prescribed fire each year cruises along at less than one-third of the need, so every year it falls further behind. British Columbia's program is tiny; it was big in the 1980s, more than a million acres in some years, but it lost favor with the public and was dropped like a hot ember. In both Canada and the United States, the Forest Services were shrunken in mindless waves of budget cutting and layoffs ever since the 1980s, and have not been restored. In 2017 and 2018, B.C. Forestry spent well over a billion dollars on putting fires out and helping the victims, and at the end of all that the government came up with just thirty million dollars for proactive fuel modification over the next two years. B.C. Forestry oversees three-fourths as much acreage as the U.S. Forest Service does, with one-tenth the number of employees. And the U.S. Forest Service itself is much reduced from its peak years.

Dozens of studies have looked at whether or not prescribed fires reduce the severity of subsequent wildfires. They do, though not perfectly each and every time. Some fires blown by hot, dry winds blast right through recent Rx fire treatments like nothing. Some scientists look at a few burns (particularly the Hayman fire in Colorado) and see treatments not working under extreme fire weather conditions and conclude, since a lot of the acreage burned in fires does burn under those conditions, that fuel treatments are almost pointless. But the preponderance of evidence does not support that view—and you do have to look at the evidence broadly, since there are such huge differences among natural forest types, as well as among fuel treatment prescriptions.

Some fires show quite the opposite: the fuel treatments make a bigger difference in the part of the fire that burns in extreme fire weather. The Wallow megafire in Arizona and the Rim megafire in California are two of those. (Megafires are simply big fires. There's no official minimum size for them. For forest fires, one hundred thousand acres works well. Grassland fires, if you even talk about them in the same breath, would need a higher bar, as they can get really vast while consuming far less biomass.)

This scenario makes sense, too: under moderate fire conditions, forest in reasonably good condition was going to burn moderately regardless of whether it had been treated, whereas in extreme fire weather it would burn severely if untreated, and moderately if treated.

In the big blow-up part of the Las Conchas fire, fire was significantly less severe where the forest had previously burned repeatedly, in both a wild-fire and a prescribed fire within the past thirty-five years. The combination of two kinds of fire had succeeded in returning the forest to an effective approximation of its natural frequent-fire condition. But where there was only a single recent burn subjected that night to extreme fire, the benefits were barely detectable. (The worst post-fire moonscapes resulted where the blow-up crossed a fifteen-year-old high-severity burn that was a still just a mess of new brush, old snags, and old dead logs.)

Fire ecologist Susan Prichard lives in Washington's Methow Valley, which bisects a national forest that has had way more than its share of fires in recent decades. Her study of the effects of fuel treatments within the 2006 Tripod megafire is a favorable one in two ways: treatments re-duced fire severity even when they burned in extreme fire conditions, and even when it had been as long as thirty years since a treatment or a wild-fire. Then came 2014 and the Carlton Complex megafire to show her that maybe Tripod's fire conditions weren't so extreme after all. Carlton was much worse. During a single nine-hour stretch that reached ninety-four degrees Fahrenheit and 12 percent relative humidity, with winds gusting over thirty miles per hour, it grew by 192 square miles. (That's the big-gest acreage I've seen for a half-day blow-up, though to be fair it was split between two side-by-side fires that later coalesced.) Past fuel treatments tempered the fires modestly but significantly that day, and more so on the later days of the fire. Thinning without burning had negligible effect, but combinations involving burns were somewhat effective.

Early in the morning of November 8, 2018, in the northern Sierra Nevada foothills, down-valley winds were howling in the canyons. That's pretty typ-ical for November. Not so typically, the fall rains so far hadn't amounted to much. That combination—fifty-mile-an-hour Jarbo winds, as they're known locally, while the hills were still terribly dry from the hottest summer on record, barely alleviated by fall rains—was all it took for ignitions from a cluster of power line failures to turn into the most deadly and destructive U.S. fire in a century. The Camp fire took only a few hours to incinerate the town of Paradise, California, along with smaller communities nearby. More than twelve thousand homes burned to their foundations, several with people inside; eighty-five people died. Human tragedies, close escapes, harrowing images, and misery in the aftermath dominated the news cycle for weeks, together with more of the same from a chaparral-and-suburbs fire that burned to the sea in Malibu at the same time.

Jim and Shannon Flanagan had opened Mamma Celeste's Gastropub and Pizzeria in Paradise just six months earlier. Jim was familiar with forest fires from his earlier years in Montana. Because of the screaming Jarbo winds and extreme fire danger levels, he was up and checking by 6:30 a.m. when the first report of a fire near Camp Creek went out on the radio. Nevertheless, he didn't expect his town to be engulfed in flames in less than three hours. At 8:20, as he was seeing Shannon off to Ponderosa Elementary with their ten-year-old, Nathaniel, a charred chunk of bark landed at his feet—a forewarning of the flaming firebrands about to arrive. Shannon and Nathaniel didn't get far before a policeman turned them back, with orders to evacuate. Within minutes the three of them were in the car crawling down the main road out of the foothills. Headlights were on as smoke clamped down on the town, turning morning into night. A block away on either side, trees were exploding. Blizzards of glowing embers blew against the windows. "We knew we were in a near-death experience. Had the wind wanted to shift and [the fire to] overrun the section of road we were on, we would have had no recourse but to get out of the car."

Jim talked to me from the temporary home offered to his family by Lindauer River Ranch, a working prune plum orchard outside of nearby Red Bluff. A previous family of refugees the ranch took in from the Carr fire in Redding had just moved out.

People who lose their homes in a hurricane "can go through something that was their home there, and find something. Here," Jim told me, "you go back, and it's a pile of ashes. The only thing that's standing is stainless steel. Everything else is gone, just melted to the ground."

Most of Paradise's population of around twenty-six thousand managed to escape despite the bouts of gridlock, and some also survived in buildings in town, but it appears that better evacuation planning could have saved lives.

Could anything have saved the town from burning down?

Not firefighters. The fire raced by leaps and bounds, literally, from its source to the edge of town, six miles in ninety minutes. The wind blew firebrands that lit countless spot fires, leaving firefighters with no unified front to attack. When the fire hit Paradise it hit in many places almost at once, all across town.

Better decisions on the part of Pacific Gas & Electric, whose power lines sparked the fire, could very likely have avoided the ignition and spared Paradise.

But if we're trying to see the broader picture, we have to assume that ignitions happen, and only some of them can be avoided. Earlier that same year, the biggest fires were ignited by a flat tire—its wheel rim banging along on highway pavement, sparking—and by sparks from a hammer striking stone. Golf clubs can do it, too.

The six miles the Camp fire crossed while charging at Paradise were brushy foothills that had burned just ten years earlier. Dried-out logs and branches killed by the earlier fire were scattered all over, ready to burn. In other parts of the West, ten-year-old burns often slow fires down or drop them to low severity; but in the Sierra foothills that effect only lasts two or three years, because the brush grows back so fast. Land ownership in this fire's path was mixed, a majority in small holdings with scattered homes, along with a few parcels of national forest and a lot of Sierra Pacific Industries commercial timber. Sierra Pacific Industries had nine-year-old pine plantations in some of the burn, but much of the burn had come back in brush. Gray pines and black oaks, the common taller species, covered portions of the area, often sparsely. An invasive weedy shrub, Scotch broom, thrived on disturbed soils such as bulldozed fire lines. When gray pine drops its bunches of ten-inch needles onto Scotch broom, they hang up all over the six-foot shrubs, creating an extremely flashy fuel. There are also swaths of sparsely vegetated serpentine soils and rock, which can often slow a fire down, but this fire went around them or jumped over them.

Applying hindsight, is there a version of thin-and-burn that could have kept the fire intensity low under these weather conditions? Ideally, yes: a savanna of grass under blue or black oaks and gray pines can do that, and is a natural vegetation type here. Recreating that savanna, given the prior history of this terrain, would have been expensive and labor-intensive, requiring planting, many years of prescribed fires, and probably some heavy-handed shrub control in places. Accomplishing that across a landscape divided among numerous landowners would be a heavy lift.

(Another natural vegetation type here is chaparral, which tends to burn severely. Chaparral taking over wouldn't save any houses or people from fire, of course, and it also may not be ecologically resilient. A top California chaparral ecologist, Jon Keeley, warns that if chaparral burns at, say, ten-year intervals rather than its historically typical hundred-year intervals, it may fail to reproduce after a few fire cycles, and may die out, most likely in favor of weedy grassland. He sees shorter fire intervals in chaparral due to warming climate and to increasing ignitions with more and more people around. The threat to resilience is greatest wherever California chaparral gets Santa Anas or similar strong down-valley winds in the fall, like the Jarbo wind at Paradise.)

In short, management of vegetation can probably help protect communities in situations like this, but for the most practical, fast-working, and confidence-inspiring ways to accomplish that in a place as flammable as California, we have to turn to fireproofing the communities themselves.

The Camp fire took by far the greatest number of both lives and homes of any U.S. or Canadian fire within memory. However, it's worth remembering the many deadly fires between 1870 and 1919. Peshtigo, Hinckley, Baudette, Cloquet, and other towns burned down, towns smaller than Paradise, but with higher casualties. Most of those fires started as forest fires in the upper Midwest states, in a time when logging of that region was intense and people were clearing land and getting rid of slash by setting fires willy-nilly, with little consideration of the risk of a blow-up. The deadliest fires of all were not forest fires: the Great Chicago Fire of 1871 and the San Francisco Earthquake fire of 1906. Today we take it for granted that a fire in one urban building won't set the entire city on fire the way the O'Learys' cow shed did. We owe that confidence to changes in urban construction materials that doubtless seemed draconian at the time. The recent spate of WUI (wildland-urban interface) fires calls for similarly major changes in house materials.

California did enact strong building codes for new construction in the WUI, effective in 2008, but they left out a number of fire-prone places, and they did not require upgrades of existing buildings. Avoiding future tragedies just like Paradise will probably require wholesale retrofitting. It's well worth doing, even if it takes community-assisted creative financing.

Another response we hear after these tragedies is that we can no longer allow development in fire-prone areas. I just don't see it (even thought I endorse similar logic when it comes to moving low-lying communities that will inevitably flood). In the first place, it is absolutely possible to build highly fire-resistant communities. In the second place, where are the non-fire-prone places in California? (Or in the Rocky Mountains?) With so many challenging political battles ahead, this one—rezoning as unbuildable three-quarters of the currently buildable attractive properties in the West—is a hornet's nest we can afford to circumvent or at least minimize, by rezoning only very selectively.

For many regions, the path toward public acceptance of more fire may lead by way of an institution called a collaborative. The thinning and burning that Ellis Margolis planned, on Glorieta Mesa, New Mexico, was the project of a collaborative funded through an early version of the CFLRP, the Collaborative Forest Landscape Restoration Program. Collaboratives are an effective process for developing local support for forest restoration—almost as effective, and more desirable, than getting walloped by fire and flooding and then seeing the light. Collaboratives are formed by groups of people living in the same region and representing different interests, like loggers, ranchers, sawmill owners, chambers of commerce, and—necessarily— environmentalist organizations. Having those activists on board is critical for assuring that the resulting projects don't get bogged down in lawsuits. The government (including the Forest Service) may have a seat at the table, but a collaborative is neither created by nor run by any government agency. The idea is that if all these people with partially conflicting points of view sit down together regularly to discuss how the nearby federal lands should be managed, they'll find they have enough in common to agree on some action plans. It seems to work.

A collaborative puts together a proposed project, details and refines it until it's basically a grant proposal, then submits it to the Forest Service for approval and funding. Local economic benefits are part of every project. Most typically, that would be timber proceeds from forest thinning. On Glorieta Mesa, the timber is so small and scattered that the thinnings were simply made available to fuelwooders, a long-standing cottage industry in New Mexico's mountains.

Its seventy-thousand-acre size made that a pretty small collaborative. The biggest one, to date, is not far away: the Four Forest Restoration Initiative encompasses nearly all of the four national forests in the highlands that bisect Arizona, called the Mogollon Rim. The total 4FRI area is 2.5 million acres, and the part of that they propose to treat is a million acres. (Pronunciation guide: "four-fry"; "Muggy-own.") Todd Schulke became a major voice in creating 4FRI a few years after he cofounded an environmentalist group, the Center for Biological Diversity. Schulke told me that Arizona's first megafire, the 2002 Rodeo-Chediski, convinced his group that the region needed forest restoration on a massive scale. That fire burned almost half a million acres, one-third of them with total mortality. In 2011 when the even bigger Wallow fire swept Mogollon Rim country, 4FRI was already underway. After a fashion.

When 4FRI won its approval there was optimism and back-slapping all around, with senators and congresspeople joining in. Eight years later, hardly anyone is happy with 4FRI. The amount of both restoration and of profitable industry produced so far fall far short of the plan. Total acreage thinned is way behind schedule and can't be sped up much because the mills and bioenergy plants needed to process the products have not been built. Schulke believes that 4FRI left the rails when the Forest Service made the wrong choice between two bidders for the first and biggest contract. The successful bidder "was obviously a scam right from the start. We knew it. We were vocal about it . . . absolutely zero chance of success." For both its lumber products and its energy products, that bidder proposed cutting-edge technologies that have yet to show commercial success anywhere near the Southwest. Sure enough, the contractor didn't get much done, kept claiming things were going great, and then went belly-up. There were only two bids for that contract, and the other one, involving an OSB plant (oriented-strand board, a plywood alternative) may have been at least as questionable in the Forest Service's view. Though that one looked reasonably promising to Schulke, he does allow that it would take longer than the initial ten-year contract for any variety of new lumber mill to achieve profitability.

There's the rub! Profitability is hard to achieve, for many reasons.

To begin with, it's almost like blood from a turnip. As Ellis Margolis told me in central New Mexico, "in this kind of low-productivity forest, you can get a money-making cut out once, and that's probably it. And they did it." A hundred years ago, or more. Here on the Mogollon Rim, the productivity is higher than where Margolis was, but still, they had a profitable timber industry here in the last century because they "high-graded," or took all the most lucrative trees they could get easily. High-graded their profits, downgraded the forest they left for later generations: smaller, crookeder, knottier trees, more density, more of the less desirable species. Turnips. To make up for it, the rules had to do a one-eighty: loggers now have to *leave* all the more lucrative trees for the sake of fire resilience and ecological health.

In the 1990s, Arizona's old growth was almost gone, and Schulke's activist group convinced the courts to halt the cutting of what remained, for the sake of endangered species. Arizona's logging industry, being based entirely on those national forests, shut down almost completely. To sell a million acres of thinning projects, 4FRI needs an industry to reemerge out of thin air. "The stakeholders thought that if you build it, they will come," I was told by Diane Vosick of the University of Northern Arizona. "In other words, if you put a large enough contract on the market, that will solve what industry needs to invest." It hasn't worked out that way, so far.

To log and mill three hundred thousand acres of trees, you need a huge investment in equipment, and investors were wary. To varying degrees, this lack of a sizable, ongoing timber industry is an obstacle to forest restoration that I hear about in many parts of the West.

Nevertheless, Todd Schulke is convinced that logging and milling could make money around here if someone would invest in modern sawmill technology designed for small trees, the way they do in Finland, or British Columbia, or Colville, Washington. They get studs out of poles whose small-end diameter is three and a half inches, exactly big enough for a stud. Most of the log trucks I saw on the roads in central B.C. should perhaps be called pole trucks—their cargo was that small. The industry in California is so stuck in the past that Craig Thomas envies other regions: "They have conferences on small wood utilization, whereas here we just think it's garbage."

Russ Vaagen of Colville, Washington, came down and built a modest-sized mill for small trees in Snowflake, Arizona, but after a few years sold it and went back north after failing to get a predictable supply of lumber without the benefit of a 4FRI contract. In Show Low, Arizona, there's a small mill managing to get by as part of 4FRI. If 4FRI can just get an experienced well-funded operator like Vaagen into the fold, there may yet be some acceleration to the snail's pace of treatments. They're working on it.

We've been discussing how to wring profits out of the low-value wood in those contracts. The question remains of what to do with an equal volume of *no*-value biomass that also has to come out of the same forests. Called slash, it comprises the limbs and tops of the lumber trees, together with some shrubs and trees too small or defective to use. Of course a number of small trees are left standing, to grow bigger, but to achieve the low density and low fuels targets that will reduce fire and mortality, restoration projects have to remove a lot of slash. Leaving *some* small branches lying around to carry a prescribed fire soon is good, but thinning in the Southwest produces way too much slash to leave all of it. It would increase fuels in the name of reducing them. The cheapest and most usual mode of slash disposal is piling and burning. You can guess some of the many things wrong with that option. Smoke. Carbon dioxide going straight into the atmosphere. Risk of the fires escaping.

From a carbon accounting standpoint, it looks appealing to get some energy out of all that waste biomass when you burn it—use it, in other words, to generate electricity or marketable heat. Surprisingly, the carbon-footprint virtues of doing so are questionable. Trucking the chips to a biomass plant uses a lot of diesel fuel unless there happens to be a plant nearby, which there rarely is. Wood debris is a lousy fuel—worse than coal—in terms of BTUs per carbon released. Beverly Law and her lab at Oregon State University calculate that in terms of carbon footprint, the best thing to do with slash is to leave it to rot in the forest. (Rotting will also release all of its carbon into the atmosphere sooner or later, but won't chalk up additional diesel fuel emissions from transport.) But she's in rainy western Oregon! In ponderosa pine country you can't leave it lying in the forest, because it's fuel, which may make the next fire a lot worse. Reducing fuel was the whole point of this exercise.

Bioenergy is also debatable on economic grounds. A biomass plant may pay a little bit for woody biomass, but probably not as much as it costs to deliver it to them. Some Asian nations are seriously looking at biomass burning as a major fuel for their electric grids, and some Arizona researchers are looking into that market for entire trainloads and shiploads of chips. But if the stuff gives off only a little more energy than it takes to truck it a hundred miles, how can it possibly pencil out if you're then shipping it five thousand miles?

In most cases it will have to be subsidized. A modest subsidy for moving biomass to a bioenergy plant, just as a way to get that fuel out of the woods, may in many cases be a good deal for society, once you figure in future reductions in firefighting costs, losses from wildfire, and losses from fire-related floods. Indeed, it was always expected that collaborative projects would need subsidies; that's why the CFLRP was enacted.

Burning slash for energy should have a relatively attractive carbon footprint once you've ruled out leaving it lying in the forest, especially if that's replacing power that would otherwise come from burning coal or natural gas (currently the case in Arizona). One reasonable way to burn it for energy is home heating with pellet stoves, at least within lightly populated areas where the smoke output doesn't rule it out. Modern pellet stoves produce more heat, with less smoke, than other ways of burning wood. Unfortunately, only the relatively solid wood parts of slash can be pelletized. The search for more good alternatives is urgent, not just in Arizona but throughout ponderosa pine country.

An alternative way of leaving it lying in the forest is to chew it up first. It's called mastication. Big machines do the chewing, of course. They have aggro names like Bull Hog, Slashbuster, Hydroax, and Brontosaurus. The Slashbuster features a four-foot disk studded with huge teeth whirling inside a partial steel bowl at the end of a long articulated arm mounted on an excavator with either rubber tires or tracks. Other styles have long drums on horizontal axes bearing huge fixed teeth or swinging knives. The operator—protected in a strong cockpit—just shoves the "head" onto limbs, shrubs, slash, or even whole trees, and the chips fly—shreds of wood up to several inches long. It's a quick-and-dirty way to prune off all the lower tree limbs while also demolishing much of the shrub layer—basically an entire fuels reduction treatment in one wild pass with just one operator.

Mastication has rapidly gained popularity. To my mind, that's because it offers fuel treatment on the cheap; but it is truly playing with fire. Masticated fuels have not yet burned in enough wildfires to provide a clear verdict. However, Bob Keane of the Fire Lab in Missoula did a broad experiment up and down the Rockies, and issued a harsh warning: if a wildfire hits masticated wood chips, they can make the fire more deadly to trees. The effect can last easily ten years after mastication, and potentially longer.

The shreds slowly compost and turn into organic soil. In arid country, that takes a long, long time. Keane's study saw very little decay of the chips over ten years; of course, they would decay faster in a mild, moist climate. In the meantime the forest floor looks and feels unnatural. If the chips are too deep, they can suppress flowers and grasses from growing, but few forests produce enough chips to raise that issue. Mastication also increases noxious weeds to some extent, probably because decomposing wood chips temporarily rob the top underlying soil layer of nitrogen.

At worst, in some fires (both wild and prescribed) the masticated-chip layer burns in ways that kills trees. Craig Allen showed me a sad research natural area in the Jemez Mountains where a prescribed fire smoldered long enough to kill all the trees by cooking their roots. In Oregon, Craig Bienz of the Nature Conservancy told me of tree-killing twenty-foot flame heights from chips. In that set of prescribed burns, all succeeded in burning at low to moderate severity except the masticated one. Nevertheless, mastication is commonly used as site prep before a prescribed fire; clearly the results can vary widely, and it should be possible to learn enough about exactly how to burn after mastication to get good results most of the time. Mastication should not be done without follow-up burning, because without that the next fire will be wildfire—and that would truly be rolling the dice.

Superior alternative ways of burning the slash for energy also exist, mostly in experimental stages. Air curtain burners are mobile incinerators that minimize smoke and air pollution, and they can be used even in the driest weather. 4FRI is looking into a variant of air curtain burners that also maximize production of charcoal—updating the old charcoal kiln. If you smother a fire—starve it of oxygen after it reaches combustion temperatures—you get charcoal, often called by a new buzzword, biochar. Simple, low-budget new ways of turning slash piles into biochar have also been invented, such as a fireproof blanket with controllable air intake. There's a market for biochar as a soil additive that increases soil's water-holding capacity. Or you can just plow it back into the forest soil. Exactly how much this benefits forest growth is unclear, but it seems worth subsidizing in any case because its carbon is extremely slow to break down. In other words, you're achieving medium-duration carbon sequestration of a substantial portion of the carbon that would otherwise have gone into the atmosphere. Alternatively, charcoal is of course a good fuel; its heat output per weight is much better than wood and better than some coal, while its polluting emissions are better than either.

A higher-tech alternative process is called fast pyrolysis, or gasification. Pyrolysis units make biochar while they also produce oil or gas that can be sold as diesel fuel or heating fuel. Mobile pyrolysis units small enough to move from one forest project to the next are on the market already, and sound ideal, since it should be more fuel-efficient to truck pyrolysis to the wood rather than trucking all the wood to the biomass burner. Unfortunately, mobile units have had reliability issues and costs so high that they might actually burn through cash faster than other methods of slash disposal, even while some costs are offset by selling the oil and the char. Pyrolysis could alternatively be performed at much bigger, regional biomass plants, with some cost savings from scale but with the added energy cost of more trucking. There are also questions about the quality of the fuel oil product. For now, pyrolysis remains another next-gen energy technology that may or may not improve enough to become worthwhile. I'll be watching.

Todd Schulke isn't the only monkeywrencher who aged into a proud veteran of a collaborative, and even a guarded enthusiast of biomass plants. In the Sierra Nevada, Craig Thomas of the Sierra Forest Legacy wrote a recent endorsement of biomass burning "in the right place, of the right size." Right size tops out at around five megawatts. "A number of ideas have been around for a while. I know one that works is a combined heat and power system. When you use the heat, you get eighty percent efficiency, not twenty-three percent. You can heat buildings, grow vegetables, dry lumber, you name it. That's where it gets sensible."

Mitch Friedman, of Bellingham, Washington, was a ringleader among EarthFirst! tree-sitters in the 1980s. By 2006 he was in a collaborative along with sawmill owner Russ Vaagen, and wrote on the Grist website that "our only hope for meeting the challenge at the scale of restoration needs is through active partnering with timber and community interests."

Rick Brown of Portland, Oregon, engaged in the ancient forest wars of the 1980s and 1990s working for National Wildlife Federation and Defenders of Wildlife. Like the three activists-turned-collaboratives above, he became an expert on forest ecology even without a PhD: he has coauthored a paper with Jerry Franklin, and Todd Schulke has coauthored with Craig Allen and Tom Swetnam. In 2009 Brown wrote an eloquent piece titled "Getting From No to Yes," about the awkward transition they each made in coming around to supporting forest restoration:

> For decades, forest activists have largely measured success by the number of things they were able to successfully say "no" to—roads, stumps, pesticides, ski developments, mines, you name it. Now, we have to learn to add a "yes" to our vocabulary occasionally, most notably the "yes" involving cutting trees in the interest of restoration. We're not finding it easy. One friend . . . exclaimed, "Man, I liked it so much better when we could just say 'no!'"

Some other naysayers of long standing have not gotten to yes. Tensions sometimes run high. "There are a few," Todd Schulke told me, "for whom my reputation is mud and that will never change . . . Still kind of a bleed-over from 'no cut.' The same people."

For Rick Brown and others who started out in western Oregon (even Jerry Franklin himself, I suspect), "simply coming to a full realization of the different dynamics . . . of Eastside old-growth forests was an essential first step. Those with 'Westside' eyes tended to look at fire-excluded, heavily ingrown dry forests and see the multilayered, multispecies [mixed-conifer] stands as classic and desirable (Westside) old growth." (Eastside means east of the Cascade crest, the forests with ponderosa pine; Westside refers to the rainy hemlock–Douglas-fir forests west of the Cascades.)

Those who oppose restoration still see infilled forests that way, to a large extent. They point to their own articles in scientific journals, articles that deny the conventional wisdom (the one described throughout this book) that the natural presettlement condition of a majority of western lower-elevation dry forest really did consist of stands maintained by frequent fire and dominated by big specimens of fire-resistant species growing at low densities. Their reconstructions claim that high-severity fires, dense stands, and fir trees were all very common and widespread in ponderosa pine range in 1870—just as common as they are today. Therefore, they conclude, reducing those elements does not constitute restoring a historically based natural condition.

Replicating the forests of 1870 is not the object! The climate has changed, and will change further. If you want to use the forests of 1870 as some kind of a guideline, the climatic warming since then requires compensating in the direction of even less density of trees, and even more dominance of the most drought-tolerant species—exactly the direction that restoration tries to take forests in.

But aside from that, the evidence for the conventional wisdom is as solid as a big old pine tree, or a whole slew of them. Big old pine trees in the high proportions consistently recorded in forestry records from 1915 or 1925 or 1940, or even today, simply could not have remained standing if high-severity fire proportions and patch sizes had always been what they are today.

Most telling of all is the ubiquity of old pine trees that bear datable scars of frequent fires. Tom Swetnam is exasperated. "The evidence is everywhere! How can they not see the evidence? You can plop me down at random in the ponderosa pine forest in the Jemez Mountains and I guarantee you within thirty minutes I'll find you a tree with ten fire scars." Together with Jerry Franklin, Hugh Safford, and a cadre of top fire ecologists, he writes detailed rebuttals of the assumptions and methods in the iconoclastic papers. And yet, he concedes, "credit the iconoclasts. They've enriched the discussion of fire ecology. There's a broad range, there are other ponderosa pine stands that sustain patchy crown fires on some scale." You could say the iconoclasts nudged the consensus from "low-severity fire regime" to "mixed-severity fire regime dominated by low-severity fires." Forest fires are patchy.

The low-severity fire skeptics are so outnumbered that in a way they remind me of climate change skeptics. Occasionally they *are* climate change skeptics.

For example, veteran activist Andy Stahl counseled me that "today's climate change focus . . . diverts attention, and blame, from the threats to forests over which we can exercise meaningful policy reform" (i.e., cutting). He suggested I take the broader view that today's change is "gradual compared to some well-known past climate disruptions, e.g., the Chicxulub asteroid impact." Really? Not to worry? Because the extinctions we cause today are not so lickety-split as the ones surrounding the terrestrial dinosaurs' demise? Small comfort, that. (And it's not a solid generalization about great extinction events, many of which left signs of rapid global warming due to excess greenhouse gases.)

I'll grant that the anti-chainsaw dogma reflects some unfortunate political realities. Back in the 1990s, proposals to protect western forests from bark beetles and fires by cutting some trees were music to timber industry ears. The George W. Bush administration's Healthy Forests Restoration Act followed, with industry support. Many environmentalists went along with it, in particular within collaboratives, striving to make sure that restoration was done right and was not just a pretext for logging. Sometimes their precautions fell short: the restoration plans were good, the trees were marked accordingly, and then logging contractors just went ahead and took the more lucrative trees, and were not penalized. For some, trust in the process collapsed. For a few, mistrust metastasized into a wholesale rejection of either forest ecology or climate change as reasons to allow chainsaws in the woods. And the industry-nostalgic voices have not gone away; you can count on both internet trolls and D.C. politicians to respond to megafires with baseless assertions blaming them on a lack of logging.

HEALTHY OR RESILIENT

The term "healthy forest" is surprisingly ambiguous. To the timber industry, it implies a forest that is growing as rapidly as possible. A wildlife biologist might insist that forest health implies a healthy ecosystem with the complete roster of presettlement animal species. Being an English major, I give it a simple and literal meaning: a forest of growing trees with relatively few life-threatening diseases or pests. Here's a more professional version of a healthy tree: "without drought stress, symptoms of pathogens or insect attack, frost damage, waterlogging, wind damage, mechanical damage, bird feeding, lightning strikes, vehicle and construction damage, chemical injury, nutrient deficiencies, dead or broken branches, loose bark, branches without leaves, oozing sap, premature mortality or abscission of parts, leaf scorch, or stem dieback."

However you define it, "healthy forest" does not actually summarize what most forest ecologists treat as the goal. It wouldn't be normal for all the trees to be healthy. Herbivores, pathogens, fires, and storms are all part of nature's workings. The ecologists' objective is a resilient forest, meaning one that can undergo a fire, an epidemic, or another disturbance and then recover to again be a forest with similar ecological functionality to what it had before—not necessarily the same species as before, but a productive forest in the biological sense, basically a measure of photosynthetic activity or of biomass creation.

Unfortunately, resilience is another highly ambiguous term: How broadly do you define the condition a resilient forest needs to be able to return to? Many ecologists require it to at least be a forest. Others don't. Be that as it may, to express what ecological forestry aims for, resilience is the state of the art.

Key thing to remember: when we say that some forests are too dense and need thinning for their health, we're basically just talking about ponderosa pine forests and their close kin, the mixed-conifer forests that include ponderosa, Jeffrey pine, sequoia, larch, and a few other fire-resistant species. Some people take the idea and run with it, applying it to any and all western forests. Timber industry spokespeople promote that oversimplification, and many senators, congresspeople, and newspaper opinion pages echo them. In truth, spruces, subalpine firs, and lodgepole pines are thin-barked, low-limbed, and more flammable; toward the end of an exceptionally long hot, dry spell, no amount of space between them will make them fire-resistant.

Thinning to create more space between them might possibly help a bit with resisting pests, but it would not mimic nature, and would be unjustifiable economically, with the possible exception of some developed areas where it protects scenery and structures.

Another argument against fuel treatments that we sometimes hear is about snags (standing dead trees) and the woodpeckers that love them. Snags from forest fires are a valuable habitat element. The argument goes that if fuel treatments are perfectly successful across the West there will be no fires but low-severity fires, resulting in a shortage of snags. In theory, that might be only a mild exaggeration, but in reality, we have a lot of mixed-severity and high-severity fire today, and no matter what we do, there's going to be plenty of it for the foreseeable future. Furthermore, the low-severity fire regime and whatever abundance of snags it maintains is the habitat the snag-dependent animals evolved with, at least within ponderosa pine forests. Snags had relatively short lives in a frequent-fire forest because they would dry out and then, when the next one of those fires came along, they would either burn up or char heavily enough to weaken and fall soon. Hugh Safford analyzed the literature and found that there were already three times more snags in the Sierra Nevada than there had been in 1850. That was as of 2014, immediately before an onslaught of new snag production courtesy of bark beetles and bad fire years. Now there must be fifty times more.

The defenders of snags also want to keep snags from being cut down in the course of post-fire salvage logging. Good point. Pro-treatment ecologists also favor leaving snags standing. California is somewhat exceptional, though, because of the way its hot Mediterranean climate favors chaparral. Dana Walsh, in planning the reforestation of parts of the King fire, marked off several burned areas for snag preservation. More broadly, though, she told me she can't plant seedlings among the snags "and expect them to live and grow to be a forest in the future. There are other places," she says, "where they can plant among the snags. In the Sierra Nevada, I have no doubt in my mind that we will have another fire in the future, and after those snags jackstraw and fall in among the seedlings, there's no chance that the seedlings would survive." They could keep the forest vulnerable to high fire severity for decades.

Both the iconoclasts and the pro-restoration scientists want to see more fire. They also agree that fuel treatments in dry forests can moderate fire and provide ecological benefits. They agree that the current rate of treatments falls far short of what is needed to usefully repair the entire dry forest ecosystem. But they diverge in how they respond to that shortfall. One side wants to hit the accelerator; the other says we're failing get enough of the forest restored, so we might as well give up on that option. Except near communities, to protect them. They're concerned that the acreage of thinning and burning that is needed will be too tough to sell to the public—yet they think that at least as great an area should be allowed to burn in wildfire instead, which will hardly be easier to sell.

When we talk about thinning forests near homes to protect the homes, there's a danger of overreaching and creating unsupportable expectations. Thinning a broad buffer zone around a community is not the most critical step in protecting housing. Wildfire's two greatest threats to houses are, first, firebrands and smaller embers that can fly from an intense fire more than a mile away—or from the house next door—and second, surface fires coming right up to the house and meeting flammable materials there. Either of those two dangers can torch homes even if you've thinned a mile-wide buffer in the surrounding forest and the thinning has done its job as designed, dropping a crown fire down to a surface fire. What can protect the house most are retrofits to the house itself, squeaky-clean fuels reductions over a small radius, and elimination of flammable materials on or next to the house exterior and other house exteriors in the neighborhood. For a buffer around a subdivision, a street and a strip of park are good. A broad buffer zone of thinned forest outside the neighborhood may provide a false sense that "the Forest Service is taking care of this, so maybe I don't have to. Not this year." That would of course increase the risk to homes.

Restoration treatments are sometimes said not to be cost-effective because any given treated area is unlikely to end up in a wildfire within the time span of the treatment's efficacy. Studies I've seen that make that point often use short or outdated datasets (too old to catch the recent bad fire years, let alone the future ones) and use skimpy assumptions on how long fuels treatments work. In Washington's 2006 Tripod fire, thirty-year-old fuels treatments reduced fire.

In one of the better studies calculating that probability, yes, the percentage of intersections (wildfires running into fuel treatments) is too low. But the conclusion the authors draw is not that we shouldn't treat, but that in order to get the benefits of the treatments and extend them further into the future, once we've treated we have to allow wildfires to come in. That's the intended benefit: to reduce the severity of subsequent wildfires. The intersections are artificially few because we haven't been allowing them. Forest Services in both the United States and Canada suppress 96 to 99 percent of their fire starts. It's a two-part proposal: a lot more fuel reduction treatments and a lot more managed fire. If both are ramped up, intersections with future fire will also ramp up—exponentially.

The goal should be to get mechanical treatments each year up to 2 percent of the dry-forest area we want to treat, and low-to-moderate severity fire (including prescribed fire and managed wildfire) up to 3 percent; then over half of all burnable forest would have one or the other within twenty-two years, and a significant portion will have had both. The most recent analysis finds that current treatments and fires meet 40 to 45 percent of that goal, in the West. You can treat those as just back-of-the-napkin numbers, but the point is that it is not an unattainable objective.

There are two somewhat contrary rationales for fuel treatments at play, and both are good, but we need to keep them straight. One of them focuses on fuel treatments as buffers around communities. The point is to make fire suppression easier—or simply possible even under extreme conditions when without them it may be impossible. These buffers are absolutely the right thing to do around communities. Some communities have enough private land for their buffer; others abut public forestland that will need treatment.

Whereas buffers may improve our fire suppression effectiveness, from an ecological point of view what we want is to improve subsequent wildfires. We'd still like to bring wildfire back to the land. Consider all the other long-term objectives of prescribed fire: to reduce the severity of subsequent wild-fires; to reduce their size and the size of their deforested patches; to reduce smoke; to reduce erosion and sedimentation of reservoirs; to increase the number of trees and the extent of forests that we end up with, and thereby to maximize carbon sequestration. For the sake of those objectives, the sec-ond rationale would place fuel treatments strategically throughout all parts of the landscape where feasible. Thinning treatments can be "anchors" that facilitate reintroducing fire. They'd be located where we would subsequently allow a managed wildfire. Next to communities? Not so much.

Once all of the constraints on both thinning and prescribed burning are totted up, it looks like a good chunk of the density reduction will have to be accomplished by wildfire.

To meet our objective of limiting wildfires, we should let more wildfires burn . . . That sounds absurd. Are we confused? Actually, that is the way it works. A hundred years of fire suppression have paradoxically led to worse fires, firestorms, megafires. (And more bark beetles.) Conversely, letting more wildfires burn *under average and better-than-average conditions*—the conditions that allow us to put the fire out if we choose to—will leave less excess fuel around to fuel those firestorms under extreme conditions. The difference between normal and extreme may be a simple as damp spring-time versus hot, dry summer or fall. The objective is to limit or reduce not the total forest area burned—that's basically impossible anyway—but the area burned at high severity within the frequent-fire-adapted forest types.

Proceeding by desultory increments since 1978, it became official U.S. For-est Service policy to let a lot of wildfires burn. Unfortunately, practice lags obstinately far behind policy. Year after year, 96 to 99 percent of fires are suppressed. Forest ecologists are tearing their hair out in frustration, trying to figure out exactly where the dysfunction lies. One paper titled "Wildfire Risk as Socioecological Pathology" describes "a destabilizing feedback loop in which spiraling fire losses are a direct consequence of policies intended to protect people and resources from wildfire."

It isn't hard to find plausible explanations for the failure to carry out the policy. When a fire on Forest Service land is first called in, the decision of how to deal with it usually falls to the district ranger. All the incentives are lined up to favor choosing the option most people will view as less risky; or where the risks are further in the future; or where the damage the fire does will be called an act of God rather than an act of the Forest Service; or simply what is expected based on the way things were done for many decades past. The ranger uses computer models to quickly sum up the risk level based on current local conditions, but even if the models say the risk is negligible, fire never quite looks risk-free. What if a managed fire does eventually get out of control and burn down some homes? That district ranger will likely go on living in the community with those victims who blame him or her. It's an intimidating prospect.

Sometimes a district ranger chooses not to suppress a fire, only to be overruled by a risk-averse politician swinging his, uh . . . his weight around.

Increasing numbers of people living in fire-vulnerable neighborhoods have grown up with the urban American assumption that fires will be put out by a taxpayer-funded agency that exists for that purpose. Naivete extends into the details. As one veteran firefighter complained, "You have houses strung all over the country and in some of the most terrible places. You can hardly drive a pickup truck into them . . . and then people wonder why their house burns down."

Neighborhoods normally seen as at risk from wildland fire are known as the wildland-urban interface, or WUI. I've seen dueling statistics on what percentage of firefighting costs go toward protecting the WUI. So let's reframe the question: Where does protecting the WUI rank among the reasons wildfires are fought today? Number one, pretty clearly. If we weren't saving homes and saving the lives of citizens and their pets, voters would surely bridle at the high costs of firefighting. It's hardly a coincidence that firefighting costs explode while the number of homes in the WUI also explodes.

There's plenty of fuzziness in the stats about forest fires burning WUI houses. First we need to clarify what's included. Most of the houses lost in western wildfires aren't actually in the woods. The fire may have originated in the woods, or more often in brushy wildland. When hundreds or thousands of homes burn down, they're probably packed close together, suburb style. For example, in Paradise, California, in 2018; Santa Rosa, California, in 2017; Black Forest, Colorado, in 2013; Waldo Canyon, Colorado, in 2012; San Diego County, California, in 2003; Los Alamos, New Mexico, in 2000; not to mention Oakland, California, in 1991, few of the burned homes adjoined real wildland, or national forest property. Most houses burn after being ignited by a flying ember; embers fly from burning trees or brush and ignite some houses, and houses pick up the torch and carry it from there.

Standard definitions of the wildland-urban interface generally include subdivision blocks like those, even if they lack trees or shrubs, as long as they are within 1.5 miles of forest—a distance picked rather arbitrarily as a maximum likely distance for a firebrand to fly on its way to setting a house on fire. (When that definition is used, those subdivisions are called the interface whereas "intermix" is used for properties that actually adjoin—or encompass—wild-growing vegetation. Firebrands are embers, only larger. We would expect long-distance spot fires to start from firebrands rather than embers because they hold heat much longer without being consumed to ash.) In fact, both forest firebrands themselves and, more importantly, house-to-house flames and embers can carry fire farther than a mile and a half from the woods.

Not only the protection of homes but the privilege of not breathing smoke is taken for granted. Some prescribed fires in California have been shut down when they were only half done, simply because too many people complained about smoke.

The good news is that opinion surveys find plenty of support (at least verbal support) among WUI residents for fuel treatments, prescribed burns, and even managed wildfire.

Andrew Larson wants us to know that wildfire without previous fuel treatments doesn't necessarily turn out badly, at least in his neck of the woods— the Montana Rocky Mountains, with its own version of mixed-conifer forest. Restoration "can be accomplished with burning alone."

He wants to present that as only a very conditional rebuttal of the dogma that says ladder fuels will make fires disastrous in today's densified forests. "I'm not gonna say we shouldn't use mechanical treatments. They can create forest conditions that will have greater resistance to high-severity fire. There's a range of conditions where you'll get restorative effects from wildfire; but under extreme conditions a fire will just burn everything up."

A quick look at Montana fires bears that out. Relative to the rest of the West, Montana has seen its share of fires and smoke in recent decades, but it hasn't had two-hundred-thousand-acre fires, and it hasn't had many five-thousand-acre high-severity burn patches—except in 1988, the year Yellowstone burned. That was absolutely a year of extreme conditions in Montana. The state's biggest historical fire, the Canyon Creek, ran twenty-one miles during a five-and-a-half-hour span, stoked by a low-altitude jet stream. This was just twenty-five miles from the benign fires Larson studies. It ended up with more than half of its 168,000 acres classed as high severity. No subsequent Montana fire can even touch its cloak.

When I first met Larson and his wife, Alina Cansler, they were two longhairs sitting on the vast floor of the convention center lobby munching from bags of snacks. They're the only forest ecologists I know who are married to forest ecologists. "It's awesome! You know, we used to talk about work at home. We still collaborate together, but pillow talk does not include science anymore. Instead we're talking about our kid, mostly."

Cleaned up now, tonsorially speaking, but still very much the ebullient mountain man, Larson teaches at the University of Montana. Seven summers in a row he led students on the continent's most coveted field research trip for grad students. They hike twenty-odd miles to the South Fork Flathead River in the Bob Marshall Wilderness. A horse packer brings in all the heavy stuff for their camp, including inflatable rafts. Every few days they float a few more miles of river to the next base camp and study site, and eventually float on out to civilization. In 2015 a fire burned the trailhead while they were in, so they had to leave the heavy stuff on a gravel bar and hike out thirty miles over Smith Creek Pass. "We kind of earned it going out." A shed full of gear burned, but their vehicles miraculously escaped.

Larson began studying this valley when it had just burned the second time in eight years. These were managed lightning fires deep in the Bob, one of the country's biggest wilderness areas. They gave him a chance to compare similar stands that were burned twice, burned once, and not burned at all since 1910. Two fires, it turns out, did a great job of restoration. Where they burned stands that started out with a high canopy of big ponderosas over infill of small lodgepole and Douglas-fir, they killed the infill trees and left most of the ponderosas thriving. An understory of native pinegrass flourished in place of low shrubs. The lodgepole pines are not reseeding themselves, because the second fire came along before the cohort following the first fire could produce cones. (That double-whammy reburn is something we worry about in regions that depend on lodgepole, but here in a ponderosa understory it's an excellent thing.)

Even more perfectly restored were nearby reburns of western larch stands. "Reburns in western larch are amazing! Picture an idealized ponderosa pine savanna." In Montana, where larches grow biggest and tallest of anywhere, their trunks actually get yellow-bellied and resemble ponderosa pine. Their foliage, though, could hardly be more different: fine, soft, and deciduous, it turns golden in fall before falling off. It stands out from the pine and spruce forest earlier in the year, too, being a brighter, paler green, accented by black horsehair lichens dripping from the high branches. Western larch is the most fire-resistant tree species in the Rockies, even more so than ponderosa pine. When I talked larch-love with Diana Six, the beetle scientist, she raved, "I wish it had a beetle so I could study it!"

(There is one little-known conifer in the West that's even more fire-resistant: big-cone Douglas-fir. Restricted to mountains near the Southern California coast, it grows with chaparral all around it, and has evolved to survive chaparral fires. It sports super-thick bark, of course. But what really sets it apart from all our other conifers is that a fire can almost completely kill its needles, buds, and fine twigs, and it comes back to life by sprouting a whole new set of twigs and needles from its scorched branches within a few years.)

Larson's advice is "Step back, and let's let nature be self-willed in these environments that society has designated as wilderness." Where it's feasible, thin before burning to reduce the risks, but where that isn't feasible, let fire burn.

As he acknowledges, we've seen plenty of examples of bad fires creating huge high-severity patches in what had once, a hundred years ago, been frequent-fire forests. He's showing us the counterexamples: fire returning to that kind of forest, with good effects despite the ladder fuels. What determines the difference?

Many factors could be involved, and probably are. But a growing number of fires of both the good and the bad kind conform to the fire suppression paradox: when we try to put all fires out, the fires that get big will be the ones we couldn't put out, because they're burning under the worst fire conditions—hot, windy, low humidity, super-dry fuels. Where we *don't* try to put fires out, the average fire will burn under average conditions, and stands a good chance of being beneficial. Some fires will still fall, by random chance, at the extreme end of the bell curve, but at least we won't be stacking the odds in that direction. And we'll be promoting patchiness, with the eventual result of making the bad patches smaller and more likely to reforest themselves.

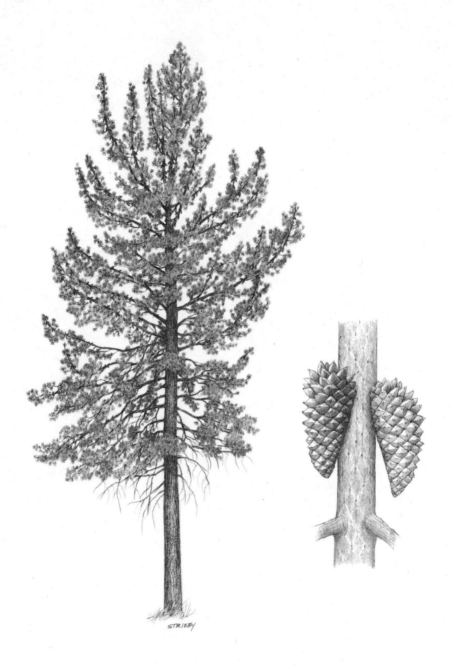

Knobcone Pine

7 NORTH, AND UP

In the not-so-distant past, climate changed rapidly. An ice age ended, temperatures warmed by more than ten degrees Fahrenheit. Tree species responded, moving hundreds of miles north, and hundreds of feet upward in elevation. They also adjusted their characteristics and tolerances. That's what trees do . . . what they *will* do in response to anthropogenic warming. With or without our help.

We know that tree species migrate with climate change. They moved northward and uphill during the rapid warming as each ice age ended, and reversed course each time the chill returned. (The Pleistocene Epoch, sometimes thought of as the Ice Age, was in fact more than thirty ice ages separated by interglacial stages about as warm and often longer-lasting than the post–Ice Age epoch, the Holocene.)

The canvas our trees move across is so vast that it can be hard to detect species migrations in real time. There's at least one clue that it's happening: for most trees in the West Coast states, the total range where seedlings are growing is slightly colder than the total range where mature trees live.

You may have heard that today's anthropogenic warming is much faster than anything today's species ever had to deal with before, but actually the science is unclear on the speed of post-glacial warming. The best proxy records of temperature are found in the ice layers of the Greenland ice cap. That ice all fell right there as snow, the layers remain in sequence going back hundreds of thousands of years, they can be dated with year-by-year resolution, and their water molecules contain oxygen atoms from the atmosphere at the time they were snowflakes. The ratio between two isotopes of oxygen reveals the temperature at that time.

Additionally, they contain air bubbles full of other atmospheric gases. These have been measured to calculate the speed of temperature change. At around twelve thousand years ago it was shockingly rapid—like sixteen degrees Fahrenheit of warming in fifty years. That's faster than current anthropogenic warming. However, those rates are for Greenland, not for the earth as a whole. Various forces came into play in different regions of the earth. Farther from the North Atlantic Ocean, temperature shifts were much slower and smaller, and in some places they went in opposite directions.

Unfortunately, where you don't have a two-hundred-thousand-year-old ice cap, you have to figure out past temperatures from other proxies that are less precise. The highest-resolution proxies in western North America are minute layers of rock that form as a stalagmite grows, in a cave. They incorporate the same two oxygen isotopes—^{16}O and ^{18}O—from the rain that fell and dripped into the cave, but the rock layers are too thin to resolve on a scale of individual years. The results so far, from stalagmites in Oregon Caves National Monument, are that Oregon experienced an abrupt warming roughly simultaneous with Greenland's at 11,700 years ago. But, equally abrupt? Probably not. (Greenland had abrupt warmings both before and after a mini–Ice Age relapse called the Younger Dryas; the Oregon stalagmite that was studied did not go back quite far enough to compare with the older warming 14,670 years ago.)

The abrupt warming 11,700 years ago brought the Northwest into a period with hotter summers than any century until the twenty-first. That lasted at least three thousand years. The winters were very cold, though, so the climate overall was not a close analogue to our century. Curiously, eastern North America didn't reach its warmest temperatures (its Holocene Optimum) until four thousand years later, probably because it took that long for the Laurentian ice sheet to melt from Ontario and Quebec. In the area from Wyoming to Arizona, some pollen studies conclude that ten thousand years ago was warmest, others that a much later millennium was warmest.

Ancient records of species composition and of fire frequency are retrieved by palynologists—scholars of pollen grains—from the undisturbed muck of small, shallow lakes. They drill a big tube down into the lake bed and pull up a cylindrical core of mud that they take back to the lab to examine. The mud settled gradually, year by year, in fine layers. Through a microscope, they identify windblown pollen—usually to genus, and in some cases to species—because pollen grains of different plants differ in shape and surface texture. Sometimes there are needles, seeds, or cones that can be identified to species, further narrowing down the pollen in that layer. Fine charcoal washes into the lakes after fires, and the palynologist measures its quantity to get a general indication of how much fire was going on at the time. From time to time, Cascade volcanoes spewed thin ash layers all across the West, and those provide instantly recognizable date markers. In between ash layers, dates are obtained by sending bits of biological material away for carbon dating. Remains of tiny animals in the muck also get identified and counted, because some of them—notably midges—are good indicators of climatic temperature.

Another excellent type of record takes the form of ancient packrat nests or middens in arid country. Preserved in heavy crusts of ancient dried packrat urine, middens can keep for forty thousand years. In some really arid places they prove easier to find than long-undisturbed lake mud. The packrat we're talking about is the bushy-tailed woodrat, usually, so picture cuter, more squirrelly critters than the usual rat. They seem to like collecting conifer needles and cone scales. Ancient middens are less ubiquitous than good lakebed deposits, so pollen studies are much more numerous than packrat studies. Each method has some advantages over the other in terms of data quality.

In any case, for around eleven thousand years the West has had climates that our tree species could have survived in their present locations—with some exceptions, which illustrate the difficulty of tree migration.

Those warm millennia of the early Holocene were a little too dry for some of today's conifers, especially in British Columbia and more northerly states. For example, ponderosa pine might have moved in while true firs had to wait, or Douglas-firs moved in but hemlocks had to wait, or the glacial-stage sagebrush tundra turned into sagebrush steppe and conifers made only minor inroads until the mid-Holocene cooling brought more rain. Yes, that is a warning to us about our century: a small shift toward warmer and drier can make the difference between pines and sagebrush.

Ponderosa pine, according to the packrat midden record, appears to have made slow and broken northward progress in the Rockies, waiting at each step along the way for the right climate to arrive. At the time of maximum glaciation, ponderosa was pushed south almost to the Mexican border. By 11,700 years ago it had made it to the southern edges of Nevada and Colorado. By 6,000 years ago it was in Wyoming, 3,800 years ago in South Dakota, and it didn't reach its final Montana mountain range, the Little Belt, until 900 years ago. Many jumps along the way were long and fast enough that they undoubtedly were assisted by pinyon jays or Clark's nutcrackers, but then the species seemed to just fill in intervening territory for a while before climate conditions were right for another jump. Sufficient rainfall in July has been proposed as the threshold it was waiting for. Because of all the delays, its net migration speed from southern Colorado to central Montana was only about six miles a century. Two-needle piñon pine also spread to most of its current range by six thousand years ago, but reached its northern and eastern limits only in the last thousand years. Its migration is ongoing.

Meanwhile, the ponderosa pines of the coastal states migrated quite a lot faster, starting from an Ice Age refugium near the south end of the Sierra Nevada.

Lodgepole pine, in contrast, was among the most successful and widespread trees near retreating glaciers in the northwest states. In British Columbia, the ice-free refugia at glacial maximum were few and small, and as far as we know lodgepole did not persist in any of them; it was pushed south into the United States. Lodgepole migrated rapidly following ice melt to fill up its present B.C. distribution. Around six thousand years ago it reached the Yukon, and there it slowed down, impeded possibly by the dense spruce forests it needed to invade. In boreal forests where lodgepole pines mix with white and black spruce, fires help lodgepole increase. Even though in theory it can handle any climate in the Yukon, it is now progressing slowly near the center of Yukon Territory.

From east-central British Columbia down into northern Idaho there's a belt of anomalously wet mountain climate—wet enough to support two hemlock species whose main range is in and west of the Cascades. Mountain hemlock had to fly 150 miles over inhospitably arid country in order to migrate from the Cascades to Idaho, and apparently that's just what it did, with lightweight winged seeds lofted into prevailing westerly winds. But if you look closely at the pollen records, it failed to take hold in Idaho until the climate cooled enough, 4,100 years ago. Even then it took hold in small numbers, starting at treeline where it was coldest and gradually infiltrating downslope. It was able to become abundant only in the last four hundred years, with the help of further cooling, the Little Ice Age. A similar disjunct population centers around Revelstoke, B.C., and apparently got there the same way—on the wind, from coastal B.C. It came even later, around 2,500 years ago. Mountain hemlocks failed to move into the mountains lying between their Idaho and B.C. sites, despite suitable habitat all the way. It's hard to say exactly why. Competition from other trees is always a factor.

At lower elevations around Revelstoke there grow rain forests of western hemlock and western redcedar. Those two species have similar immigration stories, first arriving perhaps 4,700 years ago but then, unlike mountain hemlock, doing a good job of moving into the entire range suitable for them. Redcedar didn't reach the farthest extent of its range until 2,000 years ago.

In the Southwest, conifer species were generally in the region by eleven thousand years ago. Single-leaf piñon pine took all eleven millennia to crawl about two hundred miles north to its present northwestward limit near Pyramid Lake, Nevada. In many mountain ranges, though, trees merely had to move uphill. Curiously, conifers and sagebrush sometimes traded places. During the ice ages and the Younger Dryas, sagebrush was a big part of alpine tundra above treeline. After warming allowed the treelines to move uphill, those sites are now dense subalpine forest. Meanwhile, pines were able to move out onto foothills and plains we now see as sagebrush country, with the help of annual precipitation greater than today as well as, of course, cooler temperatures. Then, in the warmer Holocene, conifer forests largely abandoned the lowlands and became a feature of the mountains. Limber pine retreated farther than others: from abounding at lower elevations of the Colorado Plateau, it was diminished to small numbers and mostly subalpine elevations. Alpine treelines moved higher than at present during the warmest early Holocene centuries. Timing of peak warmth—and highest treelines—ranged by location, from 9,000 to 5,000 years ago. By 3,500 years ago the climate was cooling and treelines moved back down to around their present elevations.

In northern parts of the West, changes were more extreme. During the Ice Age much of the region was either buried and scoured clean under glacial ice or close enough to ice that it had tundra vegetation. The ice was largely melted even from B.C. by twelve thousand years ago, and then conifer species had hundreds of miles of northward migration ahead of them. Though they generally succeeded at that, many of them have yet to occupy all of the places that appear to have a climate they would like.

For a species to move into a new area, seeds have to travel there. That limits the pace of migration. But there is a faster mode of migration in nature—*within* the existing range of any given species. Warm-adapted *genes* can travel at the speed of pollen, which is minute and easily blows long distances in the wind. For example, some Jeffrey pine cones in the northern

Sierra Nevada receive pollen blown from the Coast Range, a hotter climate, producing seeds half of whose genes are Coast Range genes. Those seeds fall to the ground along with greater numbers of seeds carrying cooler-climate genes; any seedlings that grew from them were at a severe disadvantage as soon as a late frost or an early fall frost came along, in the past. Today, however, their warmer-climate genes would give them an advantage, so that the proportion of Sierra Nevada Jeffrey pines carrying Coast Range genes will increase over time. Pollen flow tends to help populations that are at the leading edge of their species' range (the northern edge, in a warming climate) because they get lots of warm-adapted pollen from farther south. It is quite a problem for trees at the southern, or trailing, edge: they get lots of cold-adapted pollen from farther north, which will hurt their chances of survival, and never any at all from the south.

Pollen flow also goes on between pairs of species that hybridize: northeastern California will likely see increasing proportions of ponderosa pine genes in its Jeffrey pines and its hybrid pines. Eastern Oregon will likely see increasing proportions of white fir genes in its grand × white hybrid firs.

In the course of migrating at different paces, species came together in combinations that haven't been seen at any other time in the pollen record, or any other place in the world. These novel communities often combined subalpine plants (well adapted to the outgoing climate) with low-elevation plants that were just arriving, invited by the new climate. It's almost certain that we will see plenty of novel communities in the twenty-first century. This time around, there will be more party crashers arriving from other continents, thanks to the intensity of global trade. But who's to say the plant communities of the nineteenth century—the ones we think of as the God-given nature of the West—are less randomly generated than those of the early Holocene? Random chance played at least some role in their creation.

Adjusting to the Anthropocene will be more painful for our species than adjusting to the early Holocene because we are throwing so many other things at them on top of warming: fragmented and shrunken natural habitats; helter-skelter introductions of new pests, pathogens, and competitors; toxic pollutants; global deposition of nitrogen and phosphorus fertilizers.

(The term Anthropocene Epoch, which would refer to what we are in now, following the Holocene, hasn't yet been officially adopted by geological associations, but it does have good name recognition and pretty good logic: through climate chaos, humans are causing a major extinction event as well as sharp shifts in the atmosphere and the major ice sheets. There's no agreement on exactly when it began, but maybe we don't need that—after all, none of the other epochal transitions are resolved down to a single century or year.) In the Holocene, moreover, nearly all climate/soil/topography environments were were similar to ones found at the same places during the last interglacial stage, 120,000 years ago. In the Anthropocene, we're taking atmospheric carbon dioxide to levels that haven't been seen for at least twelve million years. We're creating environments beyond the limits of the evolutionary memory embedded in our trees' DNA.

Estimates are all over the map as to how much our tree species' natural migration speeds fall short of what they would need in order to keep up with their flying niche. A factor of ten is as good as any, but in reality there's going to be so much local variation and interaction that no such number is very useful.

Bottom line: during the present century, the climates that our tree species evolved with are moving north and uphill. Trees that are currently near the warm end of the range of their species will suffer from maladaptation, to varying degrees. Other trees better suited to that climate may be some distance away. Left to their own devices, they may take a long time to reach there via natural seed transport. Consequently, it might be a good idea for people to help tree species move . . .

On a mild, sunny September day in western Washington's Toutle River valley, Greg O'Neill's yellow waterproof coveralls are tattering after a few hours among the blackberry thorns. He came down from Vernon, B.C., on the dry side of the mountains, and he came prepared for wet weather, but unprepared for the way Himalayan blackberries grow down here. I'm all too familiar with that, but I'm unprepared all the same—half expecting a stroll through a rectilinear orchard, I get instead a bushwhack through what looks more like a neglected Oregon clear-cut: fireweed up over my head, big stumps, a paper-wasp nest the size of a large watermelon, huge ups and downs in the soil all hidden under broken blackberry canes vigorously objecting to our plans to explore this tree nursery.

Fortunately the weeds have in no way outcompeted the ten-year-old trees. (You can't call them saplings anymore.) A dozen tree species have shot up well past the weed competition. Lodgepole pine, paper birch, and ponderosa pine are over twenty feet tall—and those are species that don't even grow around here in the wild.

This is a special sort of plantation called a provenance trial, a side-by-side test for tree seedlings grown from different geographic sources. Look closer, and yes, there is a grid of trees hidden amid this vegetative melee: that clump of birches is perfectly square, five birches long and five birches wide, of the same dimensions as the twenty-five western redcedars standing beyond these twenty-five Alaska yellow cedars.

Greg O'Neill is in charge of this—the most ambitious set of provenance trials ever undertaken expressly to explore how forestry can adapt to climate change by speeding up the inevitable migration—north and up—of tree genes. Called AMAT (Assisted Migration Adaptation Trial), it comprises forty-eight plantation sites including this one, spread between Placerville, California, and Haines, Alaska. A majority are in British Columbia. Fifteen tree species are on trial, growing from seeds coming from a set of forty-eight source nurseries with almost as wide a range. Of his forty-eight trial sites, O'Neill says with great emphasis, Toutle is the most productive *BY FAR*. He means that these ten-year-olds are the biggest and happiest. Mild temperatures, ample rainfall, an occasional dusting of volcanic ash from Mount Saint Helens; what's not to like?

At some AMAT sites, a lot of ten-year-old conifers are so small they're hard to find. Here, the big problem is that they're crowding each other so much they'll soon have to be pruned (and the blackberry canes whacked) just to provide vantage points from which instruments can see and measure their height. O'Neill is actually eager for more tree health issues to show up, because a provenance trial isn't really getting serious about providing information until it starts showing you some limits of trees' tolerances.

O'Neill works in the Forest Improvement Branch of B.C. Forestry. (I follow the lead of locals in calling it the Forestry rather than engaging with its proper name, the British Columbia Ministry of Forests, Lands, Natural Resource Operations and Rural Development.) He explained to me the pragmatic reasons why it is easier for B.C. to shift gears than it is for other states and provinces. Timber production here is Really Big. "The Crown," meaning the province, owns all the forestry land, so decisions affecting it are all made by the one agency. For decades they've required every logged block to be replanted: their tree-planting operation is fully up to speed, on a huge scale. It was already governed by a system of vegetation zones and seed zones and a software tool for matching them, so all they had to do to start moving north, and up, was to adjust the parameters of the software.

O'Neill tells me about two key steps that enacted this paradigm shift. A 2008 revision allowed seed of most species to be moved uphill as much as five hundred meters, whereas the previous limit was three hundred meters. And in 2010 they began planting western larch in Coast Mountains valleys eight hundred kilometers northwest of the closest native range of that species.

That jump for western larch did not actually rely on a projection of future climate. The *present* climate of those coastal valleys matches the *present* climatic niche of western larch, according to exhaustively detailed climate studies. The species had simply failed to get there for whatever random reasons. It's a common phenomenon, scientifically expressed thus: "the fundamental niche of the species is broader than its realized niche." It's often said that the "realized niche" is the one that takes competition into account: i.e., a species can tolerate the climate anywhere in its fundamental niche, but in all of that area outside the realized niche, other species outcompete it, and exclude it. However, there may be niche space which the species just happens not to have migrated to yet. Barriers may have held it back. And if we are looking at the future, it's worth considering everywhere in the future fundamental niche because we don't know what the future has in store for those present competitors.

It is safe to say that, on average, the natural range of a tree species is weighted somewhat north of its twentieth-century range—certainly not in every case, but on average. Climate was fairly stable here between the years 900 and 1900; it cooled very slightly over that millennium, before turning sharply warmer. O'Neill assumes that, for practical purposes, that stable millennium gave our tree species time to do all the migrating and adapting they could do. They are attuned for life here in that millennium—and it ended more than a century ago.

B.C. Forestry now uses 1931 to 1960 as their baseline, and they project twenty years into the future. In other words, they figure out the projected future climate twenty years from now of the site they want to plant, and then look for seeds from a zone that had that same climate between 1931 and 1960. O'Neill takes great pains to drive home the point that this is such a cautious and sensible standard that you can barely call it assisted migration. "We're just putting trees back on the path that nature wants to take them on. If you plant a tree from local sources it's already two and a half degrees [Fahrenheit] out of sync because climate has warmed two and a half degrees in the last hundred years. With assisted migration as we're implementing it, you're planting a seed that's one degree warmer than the current climate of the site. The risk in planting local seeds is greater."

These trees may suffer from too warm a climate when they are forty or fifty years old, but trees are less vulnerable to climate at that age than they are as seedlings or saplings. You need them to get through their youth without getting wiped out by cold.

The great majority of climate-based transfers will only move seedlings within the present range of their species. After all, our tree species are impressively versatile. Ponderosa pine grows in places with as little as eight inches of precipitation, and with as much as ninety, with an average annual temperature hotter than sixty-one degrees Fahrenheit or colder than thirty degrees. Definitions of assisted migration often don't even include those transfers within a species' current range. A good alternate term for such transfers is assisted gene flow. Or you can say "facilitated ecological transformation" if you want to encompass transfers both within the species' range and without.

So yes, the rules O'Neill administers for the province are the utmost in caution. His personal inclinations sound more adventurous. He believes that giant sequoia, native to the Sierra Nevada in California, is already well suited to grow as a forestry tree in mountains of southern B.C. He also likes ponderosa pine for the boreal plains of northeastern B.C.: it is growing well in provenance trials there, outside Fort Saint John.

Several tree farmers in valleys of the Oregon Coast Range have been growing coast redwoods for a few decades, more than two hundred miles north of its native range. On a lot of those farms it outgrows the hometown champion, Douglas-fir. There's a terrific stand of redwoods farther north on timber industry land (location undisclosed) near my home at the forty-fifth parallel. Almost to the forty-ninth parallel, a Bay Area transplant to San Juan Island, Washington, planted redwoods for anyone who asked him for one, and they have grown well for forty years. O'Neill calls these "guerrilla provenance trials." In Prince George (fifty-fourth parallel) he admires a guerrilla sapling of giant sequoia.

But O'Neill is not the dictator of assisted migration. Tree planting in B.C. will follow data-driven guidelines. In fact, even as he is directing the AMAT, he says provenance trials are "not the gold standard" for determining where seedlings should go. He's actually wary of leaking any information from AMAT results so far, because "people might get overexcited." Early results are unreliable until the plantings have endured decades worth of local climate extremes. He thinks his computers do a better job, based on exquisitely detailed climate modeling and B.C.'s migration target, the year 1945. Matching genes with locations is not actually a simple matter of north and up. For example, narrow lakeshore valleys in B.C.'s Columbia Mountains apparently grow the perfect strain of Douglas-fir for planting *south* of there in most of the Idaho Panhandle, and even in Oregon's Blue Mountains.

When it comes to moving genes around on an industrial scale, O'Neill and all the pro-assisting scientists I know are adamant about the need for caution. If you plant a thousand acres with two hundred thousand seedlings and thirty years later you're forced to admit they're growing poorly, you've wasted a lot of somebody's money, and time. It isn't like trying a new plant in your garden, where you can water it and weed around it and coddle it for years, and if it dies it's a learning experience, not a disaster. Several attempts at moving tree genes around have indeed produced whole plantations with poor growth. O'Neill is counting on his computers to do a better job than those experiments, and for the trees to grow significantly better than the current local population would. It probably won't work perfectly every time—but then, local seeds don't either, especially in an altered climate.

Assisted gene flow hasn't really caught on in the United States as it has in Canada. "In the seventies, eighties, and nineties," I learned from Connie Millar, "we set up very narrow guidelines" because of those failed experiments. "Then in the 2000s we started overlaying climate impacts . . . There's been some loosening of those guidelines. But it's still very much fringe." She boldly advocated for assisting gene flow in a 2007 article.

Corroborating what Greg O'Neill told me about Canada versus the United States, Millar said tree planting budgets are a harsh constraint in the United States. "In Canada they still plant for timber harvest. We don't do much of that anymore in the U.S. [Forest Service]. We plant for reforestation after disturbances." Forest Service silviculturist Dana Walsh, who directs some of those reforestation plantings, explained that any climate-based shifts currently slip in through the cracks in the procurement process: if the nursery runs out of the species she needs with local provenances, but has a surplus from a hundred miles south, she can assist migration a little bit by ordering seedlings from down south. That said, she planted some giant sequoias in the King burn, encouraged by knowing the species has outliers slightly farther north. The seedlings are thriving. Four years later, she says some "are already over my head."

A few U.S. scientists have run at least one experiment supporting assisted gene flow. In the Colorado Front Range, lodgepole pine seedlings from various provenances were tried at various elevations, and the ones most likely to succeed at any elevation, even alpine, were those that originated at the lowest elevations.

Private industry replants their clear-cuts, and they do a little bit of as-sisted gene flow, but no one is tracking the big picture of what's getting planted where. And while they seem more "advanced" than the Forest Ser-vice in this regard (as well as better funded), their tree farms are behind in terms of biodiversity, clumpy spacing, variable retention harvest, leaving some snags, and minimizing herbicide use. A more sophisticated approach described by Millar: "Planting multiple species in clusters at wide and vari-able spacing would produce higher spatial heterogeneity versus the standard grid pattern, and require less maintenance (e.g., precommercial thinning) as the forest stands develop."

Millar and many others working in this field emphasize "bet hedging" by planting mixtures of seedlings of different provenances and species. "I've long recommended planting a mix, forty or fifty percent local, and the rest a mix, not just the one seed zone that we think is the future, but a mix of several, because we don't know what direction is going to be adaptive." Give natural selection a broader palette of genes to work with.

There are practical obstacles to planting mixtures. Just keeping the seedlings sorted and adhering to a complex plan will greatly slow planting crews down. "To get a contract crew to do 'random,'" Millar speculates, "you would have to give them GPS locations." Looking farther down the road, when two species are planted closely intermixed and one turns out be slower-growing in the first three decades, it may die out completely by age thirty due to competition, even if it might be the superior species over a hundred-year span.

Accordingly, a good bet-hedging strategy would be to sometimes plant the forward-looking provenances as pure small stands. If these are within the range of the species (i.e., assisted gene flow), they will be able to contrib-ute traits to the next generation over a broad area—the range of windblown pollen, as opposed to the small range of seeds.

Many conservation biologists are fiercely opposed to assisted migration. Some forest ecologists in the West share the blanket opposition, and the rest range between wary and cautiously game for some experimentation. Some wariness stems from the poor results in some forestry plantations, as mentioned above. A lot more is spillover from the adamant opposition.

I call it "fear of kudzu," and I think it's misplaced.

First off, the term "assisted migration" (AM) originally became well known as a proposed way to save endangered species (usually animals) whose native habitat is disappearing. That can be called "species rescue AM." It's quite different from "forestry AM" designed to help abundant plant species stay abundant both in their present species range and in the northward extension of it, where they will undoubtedly expand on their own in the next two hundred years, just as they have been moving north and south with climate for millions of years. (Only faster.) The idea is to sustain the ecosystem at a richer level by helping its slower-moving key members move faster than they can move on their own.

Most of the articles adamantly opposed to AM focus mainly on species rescue AM. Understandably, they cite the many examples of people—even well-intentioned scientists—translocating species with disastrous results. Importantly, the disasters have been more often microbes than animals, more often animals than plants, more often aquatic than terrestrial critters, more often small plants than trees, far more often from one continent or island to another, and hardly ever within one side of one continent. When it comes down to disastrous well-intentioned north-south transfers of forest trees within one side of one continent . . . well, I don't know of any. (Here's the closest I can find: Monterey pine on the Northern California coast is regarded by some as invasive. Old pollen deposits, however, show that it has come and gone many times from those very areas over the past hundred thousand years, including post–Ice Age times. It is properly classed as "neo-native" there.)

I don't see how the risk of becoming invasive is greater for a species that we plant northward than for species that happen to germinate northward via other paths—which will certainly include inadvertent (as well as guerilla-style) human-assisted paths even if we eschew the carefully considered ones.

Nevertheless, I've found brilliant forest ecologists on both sides of this issue. It's true that an effort at forestry AM would not have to be an eco-catastrophe to be a failure—and a failure could be large and discouraging. Malcolm North, a Forest Service research ecologist in California, told me that he favors moving seed sources (i.e., gene flow within species) but opposes moving species around intentionally, on the grounds that we don't understand forest ecology well enough. He is sobered by the way nature responded to rapid climate change ten thousand years ago with "weird combinations of species coexisting together which we would not be able to foresee." I see that as an encouraging sign—a lesson that less-than-perfect species ranges are normal and can work out fine for a millennium or two. North sees it as the path nature took, which we are now obliged to emulate, and we would just be guessing.

On the other hand, some ecologists pose our inability to accurately predict future climate as an argument in favor of AM. To insist on planting local seeds is to pick winners based on one assumption about future climate; in a sense it would be more cautious to plant a variety of provenances and species, including ones from farther south. That gives natural selection more to work with.

When you move seedlings around, you risk spreading a tree pest or disease. Probably the worst introduced-species disaster in western forests so far is white pine blister rust, which came here on seedlings of a native tree that had been exported to another continent where it picked up a disease arriving from yet another continent, and then brought it back here on nursery seedlings grown abroad. (See chapter 9.)

Our scariest emergent tree disease may be sudden oak death. That pathogen, *Phytophthora ramorum*, is classed as an oomycete—not quite a fungus and not clearly part of any other well-known kingdom. It's been found to infect a shockingly long and diverse list of plants, but on most of them it is not deadly and may not even thrive. Its chief victims so far are broadleaf evergreens of northwestern California, especially tan oaks and live oaks. Caution dictates that we rule out its current range as a provenance for any seedlings at all until we know a lot more about what plants it can infect.

Overall, the movement of pests and diseases from one part of the West to another is a serious problem. Seedling nurseries should be located just as close to where their products will be planted as they have been in the past. The risk of that kind of accident in forestry AM is surely lower than the risk in several other common practices that happen all the time, like shipping landscape plants, moving firewood, or importing products on wooden pallets from overseas. Eventually, most pests and pathogens in the West are likely to spread to all the range that suits them, and we're going to have to deal with them in some way other than hoping to quarantine them indefinitely.

For some, assisting trees in migrating is too much like "playing God." When humanity plays God, it often turns out badly. So in this case, is our primary objective to stay out of it, keep our hands clean? Or are our hands are already irredeemably dirty from playing God by burning fossil fuels, so our responsibility now is to buckle down and play God as competently as we can manage?

What exactly are the evil fates that assisted migration could avert? Trees in trouble. But specifically? The very troubles I write about throughout this book: drought stress leading to poor health leading to attacks by insects and fungi, and to a higher likelihood of dying if subjected to a low-intensity fire; an uptick in the overall mortality rate of the forest; heat stress wiping out cohorts of seedlings; slower growth.

Of course, assisted gene flow and assisted migration are not tactics to use on our standing forests, they're for the places where we want a new crop of young trees—severe burns, logged land, stands killed by insects. We expect to see more of those in the near future. The slower growth and vulnerability to mortality that we want AM to avert are in young forests and plantations, a few decades into the future.

Big mortality events of both the six-legged and the smoky varieties have a silver lining: they clear space for migration.

Neither slow growth nor mortality-thinned forests have to be seen as disasters, but they are undesirable either in terms of carbon sequestration or of timber production.

No matter what species grow on our hypothetical empty canvas, there's a risk of an across-the-board crop failure at some point in this rapidly changing climate. The risk is less total if we have a diverse mix of species. If we leave it to natural reseeding, we may or may not get a diverse mix. For burn patches that show good promise of reseeding themselves, leaving them to do that is an excellent choice, but my vote would be to salt them with a few plots here and there of seedlings from selected provenances, to make sure of more diversity, and then let natural selection go to work. If it's a large high-severity burn patch and we leave it to natural reseeding, we won't get any forest for a long time, and maybe never—we may get a brush field or grassland type that perpetuates itself by reburning. Those sites should be planted, guided by ecologists, who have learned a lot about planting for resilience. Almost anything is better than "pines in lines" (any single species planted densely and evenly in rows). Is there less risk in planting seedlings from strictly local seed sources, which reflect a climate of the past, or in planting from seed sources—and occasionally even species—matched to the expected climate?

Whitebark Pine

8 GHOSTS

From Wyoming's Wind River Range north through most of the Canadian Rockies and west to the Cascades and Sierra Nevada, tall, bleached-white "ghost trees" haunt timberlines. They lack the Christmas-tree symmetry of most conifers, instead having sloping trunks (often multiple trunks) and a few irregular branches almost as massive as the main stem. You can tell they've been dead for a long time because all the finer branches and twigs are gone. They are whitebark pines. In the cold, dry mountain air, they often remain standing for decades after the life slips out of them. The whitebark may be our most doomed pine. It is far from rare, yet by some estimates dead whitebark pines already outnumber living ones.

The West has five pine species that commonly grow at timberline. They rank as the five longest-lived species among the world's 113 pines. At least in the drier three-quarters of the West, really gnarly, ancient, wind-contorted, lonely trees at timberline are likely to be one of these five species, and the most numerous of the five is the whitebark. One in Idaho is 1,270 years old. Whitebark pine's distinctive form—a rounded broad crown featuring massive asymmetrical branches—is an integral part of timberline scenes. You can quickly conclude that a timberline tree that looks like that is a whitebark. (It could also be a limber pine; the two overlap somewhat in form and in range. The southern Rockies lack whitebarks and have limber pines at timberline; the Canadian and northern U.S. Rockies have white-barks at timberline and limber pines mostly at much lower elevations. The other three species are foxtail, Rocky Mountain bristlecone, and Great Ba-sin bristlecone.)

The whitebark is beloved, and not just for aesthetic reasons. It shelters other timberline conifer seedlings so that those species can get established, ultimately benefiting us humans via stream flow, among other paths. Its seeds are a major food source for grizzly bears and other animals. In partner-ship with one of those animals—a bird called Clark's nutcracker—it offers a remarkable case study of a symbiotic mutualism.

The scientist who figured that out came to it from the bird side. Later, as the perils to the tree became manifest, she signed on with many other scientists working to save whitebark pines. She cofounded and, for sixteen years, directed the Whitebark Ecosystem Foundation. Her name is Diana Tomback.

When I visited Dr. Tomback during research season, she was sharing a big rental cabin in Montana with her entire covey of graduate students. She cares so much about the group and their social cohesiveness that it's hard to tear her away from them even for one dinner out. Dinners at the cabin lean toward spicy, with one student hailing from Chennai and another with roots in Veracruz.

Previously, I'd seen Tomback at a conference of ecologists next to whom she looked like a silver-haired model in a fashion magazine. The other ecolo-gists looked all set to go backpacking. In truth, Tomback can backpack with the best of them, and owes her career to it.

It all began in 1970 when she was a college student trudging with a friend up into the Sierra Nevada. The Sierras have thin air and copious ver-tical relief, so most hikers stop and catch their breath a lot. As she took deep breaths under an interesting tree, a bird flew into it and began "whacking at a cone. A wilderness ranger happened by and we asked him what the tree was, and he said it was a whitebark pine and the bird was a 'pine crow, but there's probably another name for it.'" She watched the bird get seeds out of the cone. Back at the university library, she looked the bird up—Clark's Nutcracker is the name birders know it by—and found that not much was known about it; she was intrigued and eventually took it on as her PhD dissertation subject. The fact that this would require spending a lot of time high in the Sierras was not lost on her.

At that time whitebarks were not in perceptible decline in the Sierra.

The third summer of her research, 1975, was a banner year for white-bark regeneration at Tioga Pass, leading to Tomback's epiphany. Until then, she had been studying the nutcrackers simply as seed predators* that seemed to gobble up most of the pine's efforts to reproduce. She had learned plenty about how far they fly before they bury the seeds in caches to feed themselves and their nestlings through spring. She had learned to spot the bulging throat of a nutcracker holding a load of pine nuts in its stretchy sublingual pouch. They can carry more than a hundred seeds at once. She had learned to recognize the whitebark even when it is just an inch tall. And now in 1975 she saw clump after clump after clump of these newborns springing up. Why were so many of them in tight little clumps, unlike most conifer seedlings? The species is known to often grow in multitrunked form—might those simply be seedling clumps all grown up? Could the clumps be nutcracker caches that the bird never got around to eating?

---

* The biologists' verb *predate* fascinates me. It's new, as words go, with no examples pre-dating 1941 in the Oxford English Dictionary. Before then, the verb for predation was *to prey*, and the verb *to predate* simply had to do with dates preceding other dates. Well, that's a different word altogether. Biologists began using *predate*, likely by back-formation from *predation*. Perhaps they were shying away from *preying* for sounding too bloody; the OED examples of *predate* lean toward such prey as eggs, larvae, and protozoa that are neither bloody on the one hand nor vegetarian on the other. Interestingly, a few centuries earlier the preying-related words were not so red in tooth and claw, but simply meant plundering. To modern biologists, predation can be strictly vegetarian and even a vegetable can be prey: one Tomback paper calls pine nuts *prey*. But I've never heard anyone call Clark's Nutcracker a bird of prey. Hmm.

Could the bird be helping the pine by neglecting seeds, more than harming it by eating them? Could the pine species actually be dependent on this mode of dispersing seeds? Could that be the explanation for a dozen pine species having fat, rich, tasty seeds that just happen to also lack the wings most conifers put on their seeds to increase their wind-carried distance?

There she had it, the whitebark/nutcracker mutualism, in a nutshell.

The story just got better and better. Clark's nutcracker, it turns out, shares the genus *Nucifraga* ("nut-breaker") with two Eurasian nutcracker species that enjoy the same relationship with at least eight Eurasian pine species with fat, rich, wingless seeds. Those birds' pine nut diet was well known long before 1975, but the pines' dependency on birds was barely suspected before Tomback and two Europeans began reporting their studies around the same time.

Nutcracker-dispersed pines of North America include limber pines, while in our southwest, piñon pines enjoy similar relationships with both nutcrackers and pinyon jays; but the mutualism is most specialized in whitebark pines. Jays and nutcrackers are related—all in the crow family. Though Tomback and other researchers considered the similar Eurasian pines to be whitebark's closest relatives, recent DNA study calls that into question, or at least has clouded the relationship; they may have evolved the symbiosis independently.

On the other hand, once DNA analysis became available it handsomely confirmed the hypothesis about clumped stems: in only 30 percent of multitrunked whitebarks, the trunks are genetically identical, meaning they grew from one seed; in 70 percent of cases they are not, meaning they grew from separate seeds in a cache. Typically these stems show the genetic relationship of half siblings, because the seeds in the cache all came from one tree. (The nutcracker had filled its pouch while foraging on one "mother" tree, whose seeds got half their genes from pollen windblown from various "father" trees.) Multiple trunks typify whitebarks growing scattered in harsh terrain, but are uncommon where whitebarks form a closed forest. In settings with stiff competition among trees, one seedling in a clump may outcompete the others, which die. But where tree competition poses less of a survival threat than the elements do, grouping up (especially if also fusing together) might help with acquiring water and resisting wind.

Dispersal by birds offers great advantages to pines. Wind-disseminated conifer seeds lie around on the ground where more than 99 percent of them are eaten by animals that have developed excellent noses for finding them. Those that remain to germinate are exposed to the sun, which immediately sets to work overheating the new shoot. Bird-cached seeds, in contrast, are planted an inch or so deep in scattered locations, some far from trees, where hunting them down would be too inefficient even for a sharp-nosed rodent. The birds show a propensity to cache next to a rock, tree base, fallen log, or small bush—objects that protect new seedlings from sun and wind, but don't block the sun once the seedling exceeds a few inches tall.

As ancestral nutcrackers developed special adaptations for living on pine nuts—like the stretchy throat pouches, and a spectacular ability to remember cache locations—the advantages they offered to big-seeded pines created evolutionary pressure for pines to make it easier for nutcrackers—and harder for other animals—to get their seeds. These pines placed their cones on upward-reaching branches near the treetop. They ditched the seed wings. (When nutcrackers resort to eating winged pine nuts like ponderosa and Jeffrey, they laboriously de-wing them before pouching them.) Rather than dropping their seeds as the cones open, most bird-dispersed pines have ways of holding them in place under the cone scales, at least for a little while. Whitebark pines take the specialization even further: they neither open their cones nor drop them to the ground, but just hold them, all closed up, with cone scales that nutcrackers can pry apart with their beaks to extract the seeds.

Possibly the greatest advantage of getting a nutcracker to cache your seeds turns up after fires—and in warming centuries, when year-round snow is in retreat. It's the distance factor. Especially near timberline, either climate or fire can open up broad expanses of territory that trees can potentially colonize, but trees that depend on wind to disseminate them can only advance a few hundred feet per generation. (Less than that, where they have to fight a constant wind from the same direction.) They advance step by slow step, as each generation of trees at the forefront has to grow old enough to produce seeds, a delay of several decades in severe climates.

Nutcrackers, on the other hand, cache seeds as far as eighteen miles from the parent tree. Even while burdened with an added 20 percent of their body weight in pouched pine nuts, nutcrackers fly far to spread out their cache locations, reducing the likelihood of other seed eaters sniffing them out. They readily cache in big burns or in recently deglaciated terrain. They remember and relocate thousands of caches across these vast territories, often under snow that they have to excavate. Neglecting a lot of their caches (thus helping whitebarks) may not represent failures of memory so much as overzealous parenting. Tomback calculated that a Sierra Nevada nutcracker caches thirty-two thousand pine nuts, twice as many as it and its young can eat. The memory of all the cache locations begins to fade after nine months, as the next year's seeds begin to ripen.

Seed dispersal by nutcrackers is whitebark pine's key adaptation to fire. Some whitebarks grow scattered across rocky terrain where fire is unlikely to spread to them. Holding their crowns well up off the ground, they look more fire-resistant than the notoriously fire-prone subalpine firs and Engelmann spruces around them, but exactly how fire-resistant they are is debated. In some recent fires, whitebarks have actually suffered worse mortality than firs and spruces—possibly because they were already weakened by their other health issues. The effects of fire on whitebarks and their competition seem to vary. Nevertheless, there is one way in which fire unequivocally benefits whitebarks, and that is in providing openings for regeneration of new forest. When it's a big opening, nutcrackers give whitebarks a big advantage.

Having coevolved with nutcrackers, whitebark cones seem designed to give nutcrackers first dibs. Nevertheless, squirrels and bears manage to eat tons of whitebark pine nuts. The red squirrel typically nips cones off and later collects and disassembles them. Cone debris ends up in huge multiyear piles called middens. Squirrels sometimes collect unopened cones in a midden, and sometimes cache pine nuts in holes dug into older midden material. Pine nuts cached in middens rarely, if ever, grow into trees, as the midden is a poor seedbed.

Squirrels begin harvesting cones earlier in the season than nutcrackers, and their competition is a serious problem for nutcrackers. The birds have been seen fighting back by flying into a midden and flying off with an entire cone, even right in front of a squirrel. But they find most of their foraging success during years of above-average cone production, when there are more cones than the local squirrels have any use for. (Year-to-year variation in seed crops—regionally synchronized in a given tree species—is a strategy, called masting, that many tree species have evolved. Local populations of seed eaters hover around the number that the seed crop supports during the lean years. Then the tree species surprises them with a mast year and they can't multiply fast enough to eat anywhere near all of the seeds. The mast-year excess goes uneaten and provides most of the trees' future generations.) Being bigger and less mobile than nutcrackers, squirrels prefer to live where there's a variety of food resources—mixed-conifer forests, in other words. Purer whitebark pine stands are the best foraging grounds for nutcrackers because fewer squirrels live there.

In case you're struggling to picture grizzly bears up in pine trees snacking on nuts, that's not how they do it. They rob squirrel middens. Their tongues are dexterous enough to take a cone apart and separate the nuts—in their shells—but not dexterous enough to take the nut out of its shell, as nutcrackers and squirrels do. They just chomp the seeds in their shells and leave it to their digestive tracts to discard the shell bits.

Grizzly bear diets vary enormously from one region to another. Since Yellowstone's grizzlies rely heavily on whitebark pine nuts in the fall, the recent deaths of so many of Yellowstone's whitebarks of cone-bearing age stress grizzly populations, impelling bears to leave the wilderness in search of other foods, getting them into trouble with people. So far, that doesn't seem to be happening too much: the bears are finding diet substitutes without having to move very far. (For example, now that Yellowstone has wolves again, grizzlies increasingly chase them from their kills.) Still, the loss of whitebark pines undoubtedly affects birds, squirrels, and bears in ways that ramify throughout the ecosystem.

Diana Tomback is concerned that Clark's nutcrackers will abandon the symbiosis where whitebark numbers sink too low. That could make it a very long shot for whitebarks to repopulate a range—unless humans plant them.

Typically, nutcrackers visit whitebark pine areas in early summer to see how the cone crop is looking. If it is subpar, they may skip coming back that fall, and instead go off and find a better crop. (Where do they go? We can't follow them to see. Sometimes they find good whitebark crops elsewhere, and sometimes they have to settle for less preferred pine species. Limber and piñon pines are high in the pecking order; Jeffrey, ponderosa, and sugar pine nuts and Douglas-fir seeds are known backups.) It would seem to follow that if the whitebarks of any given mountain range fall below some threshold, they will lose their nutcrackers and cease to reproduce because their seeds not taken by squirrels will remain stuck inside cones. If pests or disease reduce the cone crop by half, say, we might expect squirrels to take all the cones, and nutcrackers to stay away in droves, just as they do when the normal mast-year cycle brings a cone-poor year. Some researchers report trends in this direction: below one thousand cones per hectare the birds get pretty sparse. Tomback is pretty sure this tipping point will get tipped, at least locally. So far, there is no clear case of a mountain range with no nutcracker sightings for, say, ten years in a row.

On the contrary, on sites in the southern Sierra Nevada where pine beetles inflicted worse than 60 percent mortality on whitebarks, the next few years saw a dramatic increase in whitebark seedling numbers as well as in the growth rates of the surviving small whitebarks. That's no surprise, as seedlings and small trees alike would naturally benefit from more sun where the mature trees lost their needles. But that was in the southern Sierra Nevada—the least blister-rust-infected of all whitebark pine populations.

As a tree of the highest mountain elevations, whitebark pine could fall victim to the Rapture. (That's a facetious name for the hypothesis that species forced to migrate upslope will inevitably reach the peaks of mountains, and from there depart this earth.) In more scientific terms, we say that climate envelope models show 97 percent shrinkage in the U.S. area that will have the climate in 2090 that whitebarks have lived in recently. (A lot of area opens up for them in northern British Columbia.) That would be true for most subalpine plants. On the other hand, ecologists who look at the particulars think whitebark can handle a few additional degrees of temperature in its present range. In many places it actually responds to a warmer year by growing faster, and often by engendering waves of new seedlings. It's tough, well adapted to extreme environments where it has few competitors; it covers a greater north-south range than any other North American pine; its genes are more diverse than most; it can migrate faster, notably within the large patches of high-severity fire, which we expect plenty of. Limber pines are currently demonstrating the distance advantage of bird dispersal by migrating upslope at a pace that outstrips their competition, in response to warming in some mountain ranges in the Great Basin. Swiss stone pines are pioneering in recently deglaciated terrain in the Alps, even before such terrain has any real soil. Whitebark pine has been doing the same in British Columbia.

In short, whitebark pine appears better poised for successful migration than some of its neighbor trees—if it weren't for one native pest and one non-native disease. They make its prognosis dismal.

Western White Pine

9 FADING WHITE

A spiky pair of crags named Index and Pilot lord over on an open mountain slope in Montana's Beartooths, just outside Yellowstone National Park. The slope lost nearly all of its trees in the Storm Creek fire in 1988, the summer when most of Yellowstone burned. Now it has a start on recovery as a new forest.

Diana Tomback and several graduate students are here measuring and documenting just about everything there is to measure about the growth of small whitebark pine seedlings. Before 1988, subalpine firs and whitebark pines dominated the slope. Many of the whitebark pines that survived the fire succumbed to pine beetles, but in the 1990s they were still healthy and feeding Clark's nutcrackers, which planted a decent crop of new whitebark pines here. The odds do not favor these seedlings and saplings. To grow into mature pines resembling those that burned, they will have to survive competition with faster-growing subalpine firs and Engelmann spruce, and they will have to escape white pine blister rust, a fungal disease that is often lethal. And blister rust is here.

Tomback gives us a break from the bone-rattling drive up the mountain to show me a sapling she knows to have photogenic blisters on its stem. (Pathogenic, yes, and photogenic too.) They're fat, irregular, pale orange. They produce and release spores that infect currant plants, the alternate hosts. Airborne spores from the alternate hosts initially land on pine needles and enter through the stomata, or leaf surface pores. From there, the fungus grows very slowly down through the tree, killing it ten to twenty years later by encircling the tree trunk and cutting off sap flow. Bigger trees take longer to die (the distance from needles to lower trunk is greater), but early in the process cone-bearing upper branches die, curtailing the tree's reproductive ability.

Blister rust didn't exactly come over on the *Mayflower*, but the germ of its story in North America starts even earlier. In 1605, Captain George Weymouth of the Royal Navy had his mind blown by the towering eastern white pines of New England. What he saw in them was masts. Queen Elizabeth's navy had fended off the Spanish Armada seventeen years earlier, winning the war at sea, but then launching an English Counter-Armada that was just as bad a failure as the Spanish one. Royals now saw that they needed ever larger masts for ever larger galleons. British agents cruised New England for the straightest, tallest pines and marked them as the King's property, for the King's masts. Big conifers were the key to the global arms race for most of three centuries. Eastern white pine, *Pinus strobus*, ranked as the world's greatest lumber species. Whether Europe ever grew conifers as big as New England's is unknown, but if so, they were cut down so far back in time that any memory of them had turned into tall tales. By the end of the nineteenth century, eastern white pine was nearly logged out—and holding sails up was no longer the basis of its value—but a similar and even bigger pine, the western white pine, became exportable from the Pacific Northwest thanks to new railroads. The purest, most lucrative stands were in northern Idaho. This tree, *Pinus monticola*, was declared the New Greatest: "highest commercial value of any species, wherever found." Europeans naturally wanted to grow these two species that they so admired; they had done so ever since Captain Weymouth brought seeds back, and the plantings kept increasing until the late 1800s.

Then they crashed. The American species in Europe were all but wiped out by a rust fungus that arrived from Asia in the horticulture trade. Apparently the ice ages drove the rust out of Europe after the two European white pine species coevolved with it, developing a high level of resistance. Europe abandoned any thought of growing American white pines. European nurseries suddenly had a lot of seedlings begging for a market, and saw that Maine and Idaho could use them in their clear-cuts. When a few experts warned of the risk of spreading the disease by shipping seedlings across the Atlantic, nurserymen swore that their seedlings were rust-free. In truth, rust takes a while to become visible on a small seedling. It must have been present. White pine blister rust reached New England by 1897 and British Columbia by 1921.

Americans and Canadians set to work trying to limit the contagion by attacking the shrubs that serve as the rust's alternate hosts. This fungus has a complex life history featuring five different kinds of spores. Spores from a blister on a pine cannot infect a pine; they infect an alternate host. Europeans had figured out the main outlines of this story, which features currants and gooseberries (genus *Ribes*) as the main alternate hosts. Their leaves become coated with tiny orange tubules that release the spores that infect pines. The shrubs suffer, losing most of their leaves. (Other spore generations carry the infection from *Ribes* to *Ribes* over the course of the growing season.)

In New England, currants other than cultivated blackcurrants were uncommon. Banning currant cultivation was easy, and saved many eastern white pines from the disease. After the epidemic had stabilized in the Northeast for several decades, seven states repermitted growing currants of new varieties that had been bred and tested out as immune to blister rust. Twenty years later, new strains of blister rust turned up attacking some of the formerly immune cultivars, which are now banned again.

The Northwest, with more than thirty native species of *Ribes* generously spread all over the landscape, was another story altogether. That didn't stop the U.S. Office of Blister Rust Control from declaring "War on *Ribes*." The Great Depression came along, and with it the need to create public-service jobs at subsistence wages. Soon thirteen thousand men were on WPA and CCC payrolls killing shrubs, often using herbicides whose effects on human health were poorly known at the time. Then came a real war, with a direr need for soldiers, so the war on *Ribes* took up conscripting delinquent teens and prisoners of war. Some 444 million currant and gooseberry shrubs died in the West between 1930 and 1946. After the world war ended, the *Ribes* war was deemed cost-effective enough to continue despite concerns about higher labor costs. Well, it was deemed effective by a few boosters, at least, who doctored up the appearance of a favorable balance sheet.

In 1959 the government finally threw in the towel. *Ribes* had won. Western white pine never came back to dominate the forests from which it was logged. Not only did the Office of Blister Rust Control fail to kill enough of the *Ribes* to save the pines, but we learned later that there are additional alternate hosts, wildflowers unrelated to currants: at least two species of paintbrush and at least two of lousewort.

Lest we resent Europeans for sending us their plant diseases, consider this: an American bug recently wiped out Italy's pine nut industry. The Italian cook's traditional *pignoli* come from the umbrella pine, *Pinus pinea*. (That's not the pine in symbiosis with spotted nutcrackers, which would be Swiss stone pine, *P. cembra*, of the high Alps.) These *pignoli* are much longer than most pine nuts, including piñon, whitebark, and stone pine nuts as well as the Korean and Siberian pine nuts we find in our markets. In 1999, the western conifer seed bug invaded Italy. It sucks the juice from young pine cones, leaving the seeds undernourished and too tiny to use. Just as white pine blister rust lives on European pines without doing them much harm, the conifer seed bug is not a serious pest in the wild in America. The top producer of proper *pignoli* is now Spain; time will tell whether Spain avoids catching the bug.

White pine blister rust is lethal to all American white pines, a group comprising three big forest trees—western and eastern white pine, and sugar pine—plus six higher-elevation trees—whitebark pine, limber pine, foxtail pine, two species of bristlecone pine, and southwestern white pine.

Similar native rust pathogens in the same genus attack other kinds of trees. For example, comandra blister rust (whose alternate host is *Comandra*, bastard toadflax) is widespread where lodgepole pines grow. Our pines evolved with comandra rust and tolerate it well enough; it can become a serious problem, but rarely an existential problem like white pine blister rust. Its "wave years" tend to follow periods of exceptionally high humidity with fairly cool summer temperatures.

Sugar pine suffered probably the sharpest decline of any American pine species over the last decade. Sugar pines were in decline due to blister rust already, and then, while many were weakened by rust, they were hit by the drought of 2012–2015 and the ensuing pine beetle epidemic. In 2016, Nate Stephenson found that 70 percent of his larger sugar pines were dead, a rate worse than that of any other tree in the Sierra Nevada. "Sugar pine was already on the ropes before the drought. It had been suffering decades of long-term chronic decline. It's the kind of decline," he told me, that "I call a slow-motion change. At any given year, you see some dead sugar pines, but it doesn't strike you. If thirty years' worth of sugar pine loss all happen in a single year, it would startle you and you would think there's something desperately wrong."

And then, in fact, at least thirty years' worth did die in two years, and everyone saw something was desperately wrong.

Sugar pine is still the largest pine but no longer the tallest: several ponderosa pines are taller than the tallest known sugar pine alive today.

Sugar pine cones are the largest of all conifer cones, typically at least a foot long, ranging up to almost two feet. Sugar pine cones tend to grow near the tips of high branches, like giant Christmas tree ornaments. When Clark's nutcrackers harvest pine nuts from these pendent cones, they perch upside down to reach up underneath the cone scales with their bills. (Their infamously brash cousin, the Steller's jay, likes to ram the cone feet-first and then snatch seeds in midair or on the ground.) Sugar pine "sugar" is a sugar alcohol called pinitol found in resins that exude from wounds on the bark. Native Americans used the whitish dried lumps as a sweetener, and also as a laxative. John Muir didn't mention the laxative factor when he called them "the best of sweets—better than maple sugar."

The sugar pine grows from the middle of the Oregon Cascade Mountains down through most of California's mountains and on into the mountains of Baja California. While the finest stands were in the central Sierra Nevada, sugar pines tend toward the statuesque even at the north and south extremities of their range, in Oregon and Mexico. For a giant among pines, it is surprisingly little known. Long before anyone was born who is alive today, loggers took out nearly all the best ones they could get to. The bark on mature trees is beautifully purplish but doesn't stand out quite as dramatically as a yellowbelly.

In 1826, rumors of this tree's size and its cones reached the Scottish botanist David Douglas (of Douglas-fir fame) while he was spending the winter on the Columbia River. He made a trip to the Umpqua drainage to look for the sugar pine. He considered it his best find, but one that "nearly brought his life to an end." His journal tells that the only way he could think of to collect seeds was to shoot a few cones down from high branches. His marksmanship was up to the task, but the gunfire drew eight Indians who were "armed with bows, arrows, spears of bone, and flint knives, and seemed to me anything but friendly." He cocked both of his own weapons and then faced them "endeavoring to preserve my coolness" for eight or ten minutes, at which point the subject of tobacco was brought up, via hand signs. He signed that he would trade tobacco for sugar pine cones. As soon as they left to look for some, he snuck off. That day he measured a fallen sugar pine that would rank as by far the largest pine ever measured, but the basal circumference he jotted down was so out of line with modern reality that most likely it was a slip of the pen. (It's not just the eighteen-feet-plus base diameter but the implausible taper from there to the five-foot diameter he recorded 134 feet up. Recent sugar pine champions with much slenderer bases have five-foot diameters at that height.)

Western white pine declined mainly in the 1920s and 1930s. Its number one agent of mortality was the crosscut saw. Western white constituted 20 percent of the commercial harvest volume from nine mountain states put together in 1925, and 33 percent in 1935. In dollar value the percentage would be higher, as it was the highest-priced lumber. The best western white pine forests were in northern Idaho, and it remains Idaho's state tree. But you might say the saws were racing to get there ahead of the scythe: blister rust had infected the area in the 1920s. By the time the merchantable timber was hauled away, the next-generation trees were dying, and the species had no chance of regaining its former preeminence. Five million acres were once classified as the white pine forest type. It is the dominant tree in less than one-twentieth of that area today.

Western white pines range west throughout the coastal and Cascade-Sierra mountains, where they have always been a minority species. Blister rust is well established there, diminishing southward: from Lake Tahoe south, many trees or even whole stands are still uninfected. In the Sierras white pine tends to be part of open subalpine forests, higher up than the sugar pines.

Uninfected trees can be found in most parts of its range. Presumably natural selection is at work increasing the prevalence of rust-resistant strains by culling out the more rust-susceptible lines. There seems to be little danger of the species dying out entirely, and it may be on an upward trend already.

Sugar pine and western white are similar enough that Californians often call this one "little sugar pine." The biggest specimens ever of those two species and of ponderosa and Jeffrey pine are all fairly close in size—taller and more massive than other pine species. The biggest western whites of 1920 might have won the second-place trophy, after sugar pine, but they all died and ponderosa pine took over second place.

Southwestern white pine, *P. strobiformis*, has a 1,400-mile-long range mostly in Mexico, with outposts in central Arizona and New Mexico. It has often been considered a variety of limber pine, and the two are so closely related that they hybridize where their ranges meet. Unlike limber pine, it typically stands up straight, growing mixed with other trees that stand up straight: Douglas-fir, white fir, ponderosa pine. These forests are subalpine, but not quite at timberline. Its seeds are wingless and certainly big enough to attract Clark's nutcrackers, but the cones hang down, which nutcrackers don't like. It apparently fails to entice nutcrackers strongly enough to lure them southward into its range, and when researchers looked for nutcrackers they found only nocturnal rodents caching the seeds.

Given the general scarcity of white pine blister rust this far south, the heavy infestations on southwestern white pine suggest that it is especially susceptible to the disease. Before it was infested, there was a hypothesis that white pines growing within the range of piñon pines were fairly resistant to white pine blister rust because they were exposed to a closely related native blister rust that attacks piñon pines. That rust species is rarely deadly because piñon pines have had time to develop resistance to it.

Limber pine, despite its closer genetic relationship to southwestern white, can look very much like whitebark pine. Both get gorgeously gnarly, twisted, and multitrunked. Cones are limber pine's only part that is easy to tell from whitebark: four to six inches long, initially bright green then ripening tan-brown, opening up while still on the tree, then falling and persisting all around the tree on the ground. Whitebark cones, in contrast, are rarely much more than three inches long; purplish black ripening to brown, they remain on the tree and open up just enough for a nutcracker's bill to gain a purchase.

Both grow at alpine timberlines, though in the Rockies, limber pine is common at upper timberlines only in the area south of whitebark's range, i.e., Colorado or Utah and southward. Curiously, limber pine also grows at *lower* timberlines—the lowest elevations with trees, at the feet of western mountain ranges, just above the grassland or sagebrush. In other words, it tolerates both hot and cold extremes, as well as some of the driest climates that trees can grow in in the West; and it can be found in a wide variety of environments almost anywhere in between. Since it grows slowly, it would seem to be a poor competitor; science has not yet explained how it manages to find a niche for itself in such a spectrum of competitive situations.

The oldest known limber pine is 1,670 years old, placing it thirteenth in longevity among all trees. In the genus *Pinus*, only the bristlecone/foxtail group outlives it. Both "limber" and its scientific name, *flexilis*, refer to its strikingly flexible limbs.

Limber pine is second only to whitebark pine in the preferences of Clark's nutcrackers, and it enjoys the same benefits in long-distance seeding.

It suffered widespread, severe mortality from mountain pine beetles in the recent epidemics. It is also highly susceptible to white pine blister rust. The rate of actual blister rust infection decreases from north to south. At the north end of its range, consequently, Alberta has declared it an endangered species, along with whitebark pine. So far, blister rust infections remain sparse in the Great Basin and southwest states, where limber pine is our most widespread white pine.

Foxtail pine and **bristlecone pines** are closely related, very similar species inhabiting harsh environments high in the mountains. Individuals survive for hundreds or thousands of years, rarely touched by forest fires or insect pests. One key to their longevity is their "strip-bark" growth form: old individuals usually show dead, barkless wood around much of their circumference, with the living, growing cambium and protective bark confined to a strip running up one side of the tree. Some see this as the result of being struck by lightning—a fate that becomes ever more probable as you hang around on a mountain top for thousands of years. Other scientists believe it's an adaptation to constant drying winds in an arid environment. The live strip is on the downwind side of most trees; the upwind side of the tree is impervious to harm, being already dead. That being said, these trees are susceptible to one deadly fungus, white pine blister rust. The closest thing to an exception is the Great Basin bristlecone species, which proves susceptible to blister rust in a nursery but has never been seen infected with it in the wild. (See chapter 11.)

Rocky Mountain bristlecone pine grows in the highest mountains of Colorado, New Mexico, and central Arizona.

Foxtail pines grow in two mountain areas: a short strip in northwest California and another at the southern tip of the High Sierra, with a long gap in between the two. There's some genetic isolation of the northern and southern populations; possibly they should be separated as full species. Alternatively, in the past there was interest in treating foxtails and the two bristlecone species all as one species. The biggest difference between foxtail and bristlecone pines is that foxtail grows where there's more rain, and consequently can grow bigger. The tallest one known is 118 feet, and several of the thickest ones are over eight feet in diameter. The needles are fairly short and densely, almost smoothly compressed to the branch, suggesting a fox's tail.

Since 1923, blister rust has spread almost throughout the range of five-needle white pines in western North America. The exceptions where it has not been seen are at the south end of the range in the Great Basin, the southwest states, and Mexico. Its advance southward in the Sierra Nevada has been relatively slow and spotty. At the south end, Sequoia–Kings Canyon National Parks, sugar pines and western white pines are moderately infected, but up at timberline, barely 1 percent of the whitebarks and none of the foxtail pines were infected as of 2017. These facts hint that just possibly, parts of the range are too hot or too dry for it. If true, that could become a case of a warming climate helping out with a forest health issue, at least locally. However, blister rust thrives on southwestern white pine in Arizona on bits of range even farther south and just as hot and dry as any in the Great Basin. Diana Tomback thinks that it's only a matter of time. Maybe dry climates discourage it a little, and maybe the distances between "sky island" mountain ranges slow it down. The spores have to cross tens of miles of desert to get from one piney mountain range to the next—but they're certainly capable of that, given a good wind at the same time as humid conditions. There's no absolute evidence of any climate supporting white pines that can't also support blister rust, at least in the United States or Canada.

In the Canadian Rockies and Columbia Mountains, and in the area around Glacier National Park, Montana, whitebark and limber pine mortality from blister rust is very high—it has killed more than 80 percent of the trees that grew there in 1950. Both species are listed as endangered in Canada. In the vast area in between there and the barely infected Southwest, blister rust is generally present, but has increased and killed trees relatively slowly. In fact, it is not the number one killer of whitebark pines south of Glacier National Park.

That title belongs to the mountain pine beetle. The great beetle epidemic of 1999 to 2010 began mostly in lodgepole pines and then spread into other species, becoming especially lethal to whitebark pines of the Rockies. The Greater Yellowstone Ecosystem was hit hardest: by 2010 almost half of the range of whitebark pine there had high mortality, and 99 percent had at least some beetle mortality. The few beetle-free refuges tend to be at highest elevations in the Teton, Beartooth, and Wind River ranges.

Since cold snaps (sometimes as modest as minus-two degrees Fahrenheit in spring or fall) kill pine beetles under the bark, it's easy to conclude that long periods without such cold snaps are what enabled them to advance upslope and hit whitebarks. The same explanation covers their northward advance in Canada.

However, mountain pine beetle epidemics in whitebark pines are not new. Many of the whitebark ghost trees we see are pine beetle victims from long ago: 1930 was an epidemic year. One tree killed by bark beetles back in the 1600s fell onto talus and is preserved well enough to still recognize blue-stain fungus in its wood. Looking still further back, palynologists find commingled residue of pine beetles and whitebark pines in eight-thousand-year-old sediments. That was another warm time, eight thousand years ago, the Holocene climatic optimum. The 1930 epidemic ended abruptly in 1933 with a cold winter that killed all the beetles at elevations where whitebarks live. Yellowstone's outbreak of 2007–2010 has slacked off a little, either because it's hard for the beetles to find additional victims meeting their requirements, or because October 2009 was cold enough to kill larvae.

Remember, though, that the key to a bark beetle epidemic is the beetle population explosion itself—an explosion that is only possible after reaching critical mass. In some areas whitebarks are too few for beetles to reach critical mass eating just them, but there are easily enough lodgepole pines. The epidemic developed in lodgepole pines, based on the nexus of factors described in chapter 3: fire suppression and timber management had produced region-wide cohorts of dense, mature lodgepoles. Meanwhile, warming temperatures in much of lodgepole pine range enabled mountain pine beetles to complete a generation (egg-larva-adult-egg) twice as fast as before. Those factors produced the biggest-in-recorded-history epidemic of beetles in lodgepole pine. Clouds of emerging beetles then overflowed downslope into ponderosa pine and upslope into whitebark pine. Mortality became even higher among whitebarks than among lodgepoles, possibly because whitebark genes, having less history of exposure to pine beetles, don't produce as much defense against them. You rarely see a beetle "pitched out" (stuck in a big exudation of pitch where it tried to enter the tree) on a whitebark pine.

Whitebark pines also abound in the coastal ranges—the Cascades and the Sierra Nevada. Mortality from beetles there has been spotty, with scattered foci of high death rates. In the four national parks in the Cascades, there are fewer whitebarks dead from beetles than from blister rust. Mount Rainier apparently has too few nearby lodgepole pines to transmit a mountain pine beetle outbreak. Mount Lassen, Crater Lake, and North Cascades do have enough lodgepoles; whitebarks there may have better defenses than those in the Rockies, because their winters have long been mild enough for beetle larvae. As for the Sierra Nevada, pine beetles in whitebark pine were rare before 2010, when they arrived and mounted intense but scattered attacks. We are waiting to see if the western Sierra slope beetle epidemic of 2014–2016 will move up to the peaks.

It's a warm September day high in the Pioneer Mountains in Montana. Cumulus clouds build up toward the afternoon's daily threat of showers and lightning, only to back off again. Thundershowers yesterday, not today. I'm climbing a long, gentle ridge covered with fine shards of white dolomite. Colin Maher, a wiry thirty-six-year-old with closely trimmed brown hair and a reddish beard, brought me up here to help him retrieve the last bit of data (and the equipment that gathered it) for his doctoral dissertation on whitebark pines. We drove up the mountain in his 1986 4Runner, which looks like a very clean older rig but rides and smells like a young one: it's the only car he's ever owned, and he cares for it like a concertmaster cares for her Stradivarius.

Where the highest parts of the ridge are too wind-scoured for trees, tree *species* can still creep up pretty close to the ridgeline as long as they remain shrubs, huddling in continuous low thickets. The word for tree species growing in this alpine shrub form is krummholz, "crooked wood" in German. The trick is that thickets catch snow so that it doesn't blow off the mountaintop. To get a deep winter-long snowpack, you need krummholz— and vice versa. The krummholz thicket shows you the exact dimensions of the snowpack. Any branches that grow straight up from the top of the krummholz die because they stick up out of the snow and get sandblasted by windblown bits of ice.

The ice-blasted explanation for krummholz has been around for decades, but Maher's experiment is the first to prove it, while disproving the alternative explanation—that cold temperatures themselves cause trees to grow as krummholz. He took horizontal branches at the top of the krummholz and tied them, upright and protruding, to stakes. He put mesh cages around some of these erect branches, protecting them from driven snow but not from cold, and he wrapped cylinders of clear plexiglass around some of them to warm them in the summer. Erect branches in the wind died, while wind-protected ones were fine. He concludes that "a warming climate may not lead to treeline advance at krummholz treelines unless winter wind and snow patterns also change."

Nearly all of the krummholz on this ridge is whitebark pine, which tolerates cold a little better than its competitors, subalpine fir and Engelmann spruce.

(Actually, a krummholz shrub does occasionally succeed in growing a tree. Once the vertical stem gets up there a gap may develop between the needles on the small tree and those on its krummholz skirts, because the small branches immediately above the snowpack surface may get sand-blasted away while the central stem survives; several inches above the snow surface the scouring is less intense, and branches and needles survive.)

The ridge we're on now has old, bleached tree trunks lying around in the open space above the highest krummholz, suggesting that centuries ago the slow advance of krummholz actually produced a whitebark pine forest, which eventually died in some major setback—a fire or a beetle epidemic. Maher saws cookies and drills cores from some of the trees, checking out this intriguing phenomenon in case he decides to turn it into another research project someday.

Spending time in krummholz has given Maher hope for whitebark pine's future. Pine beetles inflicted terrible mortality on the whitebark pines of the Pioneer Mountains, but they left the krummholz alone. That's typical: they leave four-inch-diameter trees alone because they wouldn't find enough food in them. But unlike four-inch trees down in a forest, four-inch krummholz pines bear lots of cones in some years. They're sure old enough, and they're sure getting plenty of sunlight. On this ridge today, we're seeing tons of cones, decent numbers of Clark's nutcrackers, and abundant whitebark pine seedlings just a few years old—probably from the mast years of 2005 and 2006. It's now 2018, probably the best year for whitebark cone production since those years. Maher hypothesizes that areas of whitebark krummholz may continue to escape pine beetles and provide seeds to bring populations back after epidemics.

(Diana Tomback, based on her geographically broader studies of treelines with whitebarks, is more pessimistic. She finds whitebark seedlings—and viable seeds—to be generally scarce at treelines.)

The abundance of cones inspired Maher to take a few home last spring, quarter them with an ax, and infuse them in spirits. This is actually not as weird as, say, brewing bark-beetle-fungus beer. At some point he happened to learn that Italians use almost every aromatic plant native to their land as a flavoring for a traditional liqueur. He knew about stone pines—the Alpine trees with a nutcracker bird as a symbiotic partner—and he had noticed that the unhardened young cones on whitebark pines are fabulously fragrant, so he put it all together, googled "pinus cembra liqueur," and found recipes and examples from each country in the Alps. One named Zirbenz, quaffed at ski resort bars worldwide, has been made commercially in Austria since 1797.

A few miles from Maher's krummholz and a thousand feet lower, Diana Six had shown me telltale evidence, a small tree "flagged" by a debarked section of eighteen inches of trunk, ten feet up. Trees often respond to blister rust by loading up their phloem with extra sugars. Red squirrels detect the sugar and gobble it up, bark and all. Yes, blister rust is in the Pioneers. It hasn't hit them as hard as many other Montana mountains. She tells me rust is much more prevalent in the wetter mountain ranges, and this is a drier one.

Mortality from either beetles or fire may help natural selection go to work on whitebark pines, if the mortality leads to a big new generation of whitebarks. But that's a big *if.* No matter how many cones we find to celebrate, the plight of whitebark pines remains a truly wicked problem. They face not one but three existential threats: beetles, blister rust, and drought stress. Natural selection (and human efforts, as we will see in the next chapter) should theoretically help future populations of the tree adapt to handle any one of those threats, but any such improvements locally can get nipped in the bud by one of the other threats, defeating natural selection. Then on top of that, whitebarks have this wonderful partnership with a bird, but they are dependent on it, and the partner may abandon them if their populations fall too low.

The departure of nutcrackers would pose a bitter irony in areas where white pine blister rust is the chief mortality agent. (That's the pattern from Montana's Glacier National Park north through the Canadian Rockies: terrible rust mortality, not so much pine beetle.) The pines that survive are likely to have some degree of genetic resistance to blister rust; they are the very trees whose genes we most want the nutcrackers to disseminate. But if they are too few, nutcrackers may pass them up. This mutualism—a superb showpiece of natural selection—could fail the whitebarks and defeat natural selection's ability to perpetuate them.

Foxtail Pine

10 RESISTANCE

The Dorena Genetic Resources Center, in the Oregon Cascade foothills, looks like a sort of raised-bed nursery for conifers, with macabre over-tones. One garage-like building is devoted to infecting a crop of year-old seedlings with spores of a lethal disease. They are wheeled in by the boxful, the doors close, a fog machine raises the humidity to 100 percent while air conditioners chill the room to sixty-two degrees Fahrenheit, and racks of currant leaves are suspended over them. Invisible clouds of basidiospores drift down from creepy-looking tiny orange tubular telial growths that coat the currant leaf undersides. I almost expect to be given a white suit to wear before entering this building. But no, no diseases communicable to humans are present; it's just white pine blister rust.

Outside in the sun, in row upon row of boxes, older saplings gradually die—or live. Of the eight western species of white pine (all planted here), the boxes of foxtail pine stand out with their near-100-percent death rate. The species farthest from home, southwestern white pine, surprisingly looks happiest in the rainy Oregon climate; at least it grows the fastest, and its beautiful whitish-bluish needles stand out. Dense coppices of eighteen-foot trees all somehow spring from crowded boxes of soil mix just ten inches deep. "We could almost harvest them," says my host, geneticist Richard Sniezko, a wiry man with intense blue eyes and rapid-fire speech accented with hypermobile facial expressions.

He loves to show off whitebark pine boxes. Most contain one or two rows of robust green ten- or twelve-inch seedlings, flanked by many rows each displaying a different mode or different pace of deformity and death. Each row is one "family," meaning seeds all from one parent tree. All of the parent trees were preselected: they survived in the wild in communities where a lot of whitebarks typically had died of blister rust. A tree whose seeds produced rows of dead seedlings must be an "escape," meaning that rather than having great genes, it just got lucky. So far. Many are called, but few are chosen. Each happy family surviving in this plague camp shows that its parent tree passes the test and is part of the Resistance. In years with good whitebark pine cone crops, technicians return to the parent tree to collect seeds. In consideration that pine beetles or a fire may kill the parent tree before the next good cone crop year comes along, technicians may go sooner and clip some branches to graft and grow in a seed orchard—a nursery with a different role than this one. (Another hamlet hosting trials-by-fungal-infection—in an area rife with especially virulent strains of blister rust—is named Happy Camp. It almost sounds like locating trials in Happy Camp was a case of someone reading too much Orwell.)

The object is to confirm which individual trees carry genes for resistance, and then to use their seeds to grow nursery seedlings for replanting in the wild. Over time, nature selects disease-resistant genes. That's how evolution by natural selection works: at some point in past evolution, white pine blister rust killed off the strains of European and Asian white pines that did not resist it or tolerate it, leaving those continents with white pine populations rich in resistant or tolerant plants. (A tolerant plant is one that can live with the fungus: it develops infection cankers, but eventually succeeds in healing them, whereas a resistant plant seemingly walls off the fungus either within the pine needles or around a few stem cankers.) North American white pine trees randomly have genes that give them various types and degrees of resistance, but since blister rust is not native here, those genes were never selected for, and most individuals lack them. After blister rust arrived on our shores and attacked three of our very best commercial lumber species, foresters didn't want to just wait for natural selection to bring the white pines back from the brink of extinction—they wanted to plant resistant seedlings, so they established breeding programs. Dorena has been working with breeding western white pine since 1966.

Whitebark pine and its high-elevation kin are not commercial lumber species, but they are ecologically and scenically valuable (as well as including the world's oldest trees). There are reasons to doubt that natural selection can save them. With whitebark, there is the danger of local populations falling below the threshold for attracting Clark's nutcrackers, as discussed in chapter 8. Worse, when bark beetles, fire, and blister rust are all attacking a species that often tends to be sparse, the naturally occurring uninfected "plus trees" may be too few and far between to pollinate each other. If a plus tree is pollinated entirely by anonymous pollen donors, its seeds will have rather weak resistance because many of those pollen donors are lucky escapes rather than resistant trees. Natural selection may be ineffective when only a modest percentage of seedlings have two resistant parents. If we could plant colonies of seedlings from a large number of resistant parents here and there throughout the range where whitebark pine will be able to live in the future climate, the resistant genes will be able to spread much faster.

For now, screenings in Dorena serve only to separate survivor trees into actual rust-resistant trees and lucky escape trees. Seedlings from the former will survive their trials at about a 50 percent rate, whereas seedlings of the latter will have lower survival rates. The seedlings grown here, even the healthiest ones, won't get transplanted. Sniezko's team tells the Forest Service which parent trees are resistant, and the Forest Service can plant other seeds from the same tree—either in the wild or in seedling orchards—and expect up to 50 percent survival. They may have more seeds in storage from those trees, or they can use GPS to relocate the trees and collect more seeds.

Collecting whitebark seeds from plus trees is no cakewalk. Carl Seielstad, a professor at the University of Montana, previously had several jobs in the woods. In one of them he was sent out to locate certain trees three times. The first time was to climb them when the cones were immature and attach cages of woven wire mesh to protect the cones from squirrels and nutcrackers for long enough for them to mature and produce viable seeds. The second time was of course to retrieve the matured cones, and their cages. And the third time was to protect the entire tree when a wildfire threatened it. His instructions were to cut a fire break around the tree, but not a hundred-foot clear-cut—just enough to make the tree safe, and no more.

Of the plus-tree progeny that survive in the seedling orchards, some can be planted out as seedlings, and some can be grown to cone-bearing age in the orchard and get pollinated there by an equally resistant parent. That will be several decades from now. Two ways to speed the process up are at least theoretically possible, and are being attempted. One would be to take pollen cones from one wild-growing resistant tree and use them to hand-pollinate young female cones on a nearby resistant tree. The other is to cut cone-bearing branches from a resistant tree and graft them onto whitebark saplings in an orchard. The grafted trees should reach cone-bearing age (and produce two-resistant-parent seeds) faster than normal orchard trees, but it's hard to promise exactly how much faster.

When foresters plant whitebark pine seeds or seedlings in the wild, they often plant them in clumps, mimicking the way Clark's nutcrackers do it.

While white pines and blister rust get most of the attention, the Dorena Center has worked on other diseases too. It chalked up a great W by breeding resistant strains of Port Orford cedar in relatively few years of work. You may not have heard of Port Orford cedar, as it has a small native range down by where Oregon, California, and the Pacific meet; but it ranks as Britain's favorite decorative evergreen (under the name Lawson cypress), this writer's favorite fragrance, and Japan's favorite lumber, if we judge by the price they pay for it. A factory in southern Oregon makes high-end arrow shafts out of it. Hundreds of named cultivars as well as the native tree are lethally vulnerable to the invasive root pathogen *Phytophthora lateralis*, part of the nefarious *Phytophthora* clan that causes Sudden Oak Death and Irish potato blight. Dorena is already supplying *Phytophthora*-resistant seedlings for re-forestation, for arboreta, and for plant breeders who are grafting ornamental Lawson cultivars onto them.

There go those government scientists again, doing valuable work for the private sector to profit from.

The summer of 2017 brought several huge fires; the biggest in the United States was Oregon's Chetco Bar fire. It was eclipsed that summer by fires in British Columbia, and then also eclipsed in December by California's Thomas fire. Dana Walsh, the silviculturist leading the planning for restoring forests in that burn, sees it as a great opportunity to plant *Phytophthora*-resistant cedar seedlings as well as blister-rust-resistant western white pine seedlings.

However, resistance to white pine blister rust looks to be a tougher nut to crack. These trees live for centuries, and we want the rust-resistant seedlings to live for centuries. The problem is that blister rust evolves too; it evolves virulence that can overcome major gene resistance in a pine. The genes for this virulence already exist and have shown up in several locales in the West, killing seedlings known to have major gene resistance. So "major gene resistance" is paradoxical: it sounds strong, in that it can completely block infection, but because just a single gene carries it, it can be easy for the evolving pathogen to defeat through natural selection of its own.

What Richard Sniezko would much rather find is called, paradoxically, partial resistance, or tolerance, or slow rusting. It lives on several genes and works in several independent ways. None of them can cure blister rust, but collectively they give a pine a decent chance at a long, albeit scarred, life. Sniezko's job demands that he spend years searching for the best combination of resistance genes he can find, using his seedling trials and his intuition to pick the best. He'll send those seedlings out, saying, "these are the ones to plant," and neither he nor anyone he works with will be alive in a hundred years to see how they did after the fungus got a chance to adapt to them. At twenty-five years old, the earliest field trials to date offer grounds for optimism.

That's life in the battle to protect our conifers from exotic fungal disease.

A major resistance gene has been identified in four of the West's white pines. Whitebark is not one of the four, but it has a higher percentage of trees with partial resistance than any of the four. One corner of its range is a hotbed of resistance: "In Mount Rainier National Park," Sniezko tells me, "almost every tree we tested is resistant." There are other places, like Sequoia National Park, with little or no blister rust on their whitebarks, but that seems to be because blister rust just hasn't spread that far yet. Those pines may or may not be resistant. At Mount Rainier, in contrast, blister rust has been killing a few white pines for decades. The resistant whitebarks Sniezko tested are from cones on trees that survived, thus preselecting themselves. His comment means that hardly any of the survivors turned out to have been just lucky escapes: good numbers of their offspring survive in Dorena.

A few scientists are looking into other approaches to saving our pines from rust. Genetic engineering, for one: find the genes that impart tolerance in European pines, splice them into American white pines, and grow nursery stock from those embryos. A more obscure approach would utilize fungal endophytes. Scientists have been learning—only quite recently!—that microscopic symbiotic fungi live inside most plant leaves. (Endophyte means "inside a plant.") These fungi are naturally interested in fending off potentially competing fungi, including blister rust. Some endophytic fungi in white pine needles have been identified and correlated with higher levels of rust resistance. To use them, we would have to confirm those correlations and also learn how best to inoculate fungal endophytes into pine trees. Very little is known about how they get around, in nature, though obviously they do somehow, and potentially they could spread from pine to pine once they are in an area. Gene splicing and endophyte inoculation are both extremely challenging technologies for combating blister rust, and research has barely begun. Even though in theory they could act quickly, they are unlikely to beat old-fashioned selective breeding to the finish line.

With western white pine and sugar pine, the breeding program has fifty years under its belt. Seedlings with two apparently resistant parents are available, and have been the ones planted for the last decade. Unfortunately, the quality of resistance is lower in those trees than in whitebark pine, because it's more prone to virulence developing in the fungus. The results look decent so far: in six field trials around Washington planted with hundreds of western white pine seedlings twelve years ago, the blister rust mortality rate after seven years was less than 1 percent, and the rate of visible infection, though much higher, is only a fifth as high as in the susceptible (control) seedlings in the same orchard. But with even the controls doing that well, a big factor in the benign result is clearly that blister rust hasn't enjoyed favorable conditions in this decade. It also hasn't had time for evolution to select its counterweapon—genetic virulence that overcomes the major gene resistance. If, say, 30 percent of the purportedly resistant seedlings survive for fifty years even in the most rust-ravaged areas, Richard Sniezko would call that a good result.

Great Basin Bristlecone Pine

11   THE ENDURING

In Nevada, California, and Utah, near the summits of scattered mountain ranges, grow pines that have survived everything the environment has thrown at them for five thousand years.

They're called *Pinus longaeva*, Great Basin bristlecone pine.

Will they survive the next two hundred years? Does their incontrovertible endurance attest to unexpected powers of survival?

It's a bright spring morning at ten thousand feet above sea level, thirty-two degrees Fahrenheit, thin air. Many of the bristlecone pines are half dead. Or three-quarters dead, or 95 percent dead. Indeed, it's safe to assume that virtually all of the community's elders—the two-millennium-plus club—are showing a lot of bark-naked dead tree trunk.

And do they ever show it! Not the subtle silvery gray you expect (well, maybe a little of that) but great streaky smears of russet, golden ochre, cream, and russet-tinged sooty blackish gray, all textured with fine-patterned ridges, and all as hard as stone. A fingernail can't dent it. With the grain of the tree trunks exposed, you see all the literal curls and loops and contortions inevitably inflicted on individuals that have spent every waking minute and every dormant minute on a windy mountaintop ever since the days of pharaohs so early that they built their tombs of mud and sticks.

I think these are the most gorgeous trees I ever walked among.

If you think just because they are half dead that they aren't getting along just fine, you are dead wrong. You're looking at the world's number one longevity strategy, working!

Bristlecone pines live in places that endure drought and cold for most portions of most years. Occasionally, over the centuries, they respond to worse-than-usual drought stress by allowing vertical sections of their cambium—the growth-generating part of the tree trunk, just under the bark—to die. With less living tissue to support than they had before, they survive the drought. The dead side of a bristlecone tends to be the windward side, suggesting other possible explanations. It could be that a root dies after erosion exposes it, and then the cambium connected to that root dies. (In bristlecone pine, sap flows up and down in vertical sectors of the tree and is blocked off from flowing circumferentially to other sectors. This sectored architecture is seen in only a few tree species.) Anyway, the dead part of the trunk is still doing its job perfectly—namely, holding the tree up. Just like any other tree trunk, only better. The heartwood of any tree is made of dead cells, and serves simply to hold the tree up. In most species it fails at the job sooner or later, succumbing to rot fungi. The dead-looking trunks of bristlecone pines do the job better (longer) than any other kind of tree, because they are so dense, dry, and resinous that they are immune to rot. Their most common cause of total death is lightning. But even then, after bristlecone pines actually die all the way dead, they can still persist, standing or lying on the ground, for thousands of years more.

(By finding and sectioning old logs or even small chunks of bristlecone wood lying around, dendrochronologists have put together a continuous 8,837-year tree ring record. By crossdating, or matching a sequence of rings in the older part of one with a sequence in the younger part of another, they can ladder different trees together to make a tree-ring record far longer than any one tree's life. It was this extended bristlecone pine record that first calibrated the carbon-14 dating technique, back in the 1960s.)

Partial dieback is a strategy many plants employ to deal with shortages of water, or of light. Most often they jettison a portion of their foliage. Giant sequoias, for example, have been sacrificing a lot of their needles because of California's severe drought. In some kinds of trees, multiple entire limbs may be allowed to die. In an unusually extreme variant that goes by the term "strip-bark," some arid-country pines and junipers let vertical sections of cambium die. Where the cambium has died, the bark covering it falls off. A bristlecone may start by losing a strip of cambium just six or eight inches wide, leaving enough living cambium to still support limbs on all sides of the tree. Over time, more and more strips die. Eventually only a strip of cambium may survive, maybe a foot wide on a trunk many feet in circumference, and you can follow the live strip upward to a narrow branch or subtrunk of living tree.

In recent years, contrarians quibble that a creosote bush ring, or maybe an aspen clone, may be the oldest living plant—neither of them with any single stem more than a few hundred years old. Give me a break! If looking at a five-thousand-year-old *tree* can't shut you up, I don't know what will.

Now, I can't actually say that I saw a five-thousand-year-old tree. I saw the forest that includes at least one five-thousand-year-old tree. From 1958 to 2012, the official oldest tree was Methuselah, 4,851 years old as of 2019. Edmund Schulman, an assistant to A. E. Douglass, the first dendrochronologist (chapter 4), cored it and Tom Harlan made the calculations. (An ancient bristlecone might not have a complete set of rings in any one radius from the pith—and only an actual druid would be able to locate the pith with a corer anyway—so calculating the age using multiple cores and cross-dating takes endless time and effort.) The Forest Service wisely removed the sign on Methuselah, fearing souvenir-cutters if not psychopathic vandals. In 2013, Harlan announced that another tree Schulman cored in the 1950s, still healthy after waiting decades to have its age figured out, turned out to be 5,067 years old (as of 2019). Ratcheting up the secrecy, Harlan swore he would take the secret of its location with him to the grave. He has done so, and also left Lady Older-than-Methuselah nameless. There are doubtless others older than five thousand years. Given the number of bristlecones growing atop the White Mountains, I strongly doubt that the oldest one has been cored.

A man named Donald Currey had been an object of popular scorn ever since he had the Forest Service cut a bristlecone down so that he could determine its age—4,844 years old in that year, 1964. It was the world's oldest living tree up until the moment it was felled. He was a graduate student doing dendrochronology work when two of his corers broke off in the dense, gnarly wood. He applied for and was granted permission to fell that one tree—even though it was distinctive enough to already be known by a name, Prometheus. Currey unfortunately did not live to hear Harlan's announcement of an older tree.

Prometheus lived in Nevada, in what is now Great Basin National Park. The range of the species consists of an archipelago of "sky islands"—the tops of fifty or so mountain ranges of Utah, Nevada, and easternmost California. Similar habitat in Colorado, northern New Mexico, and one mountain in Arizona hosts Rocky Mountain bristlecone pine, which used to be considered part of the same species.

Though bristlecone pines grow mixed with other species at slightly lower elevations, the longevity strategy seems to work best at treeline, where their growth is slowest and where there are no other tree species to compete with. Too cold and too dry, on poor soil. Here in the White Mountains the soil's parent material is dolomite, which makes the soil quite alkaline—another stressor that many trees can't handle. Lack of water enforces wide spacing between individual trees, and that in turn means there's no advantage in growing either fast or tall, ever, to compete for sunlight. By adapting superlatively to drought and cold, bristlecones liberated themselves from having to put much energy into growth.

Maybe that could have given them a decent niche even with run-of-the-mill longevity. However, slow growth and drought tolerance naturally led to dense, dry wood that resists rot, and a tree that doesn't rot is already a good part of the way toward long life. (Rot fungi are pretty ubiquitous on a bristlecone's dead wood surfaces—in fact, they cause all of those gorgeous warm colors—yet the resistant wood does a great job of restricting them to the surface.) The next part of the way is avoiding fire, and that's also taken care of by the arid habitat: fires don't spread here, because it's too dry for the trees to grow close to each other, and too dry for a continuous understory. Fires do reach and sometimes kill bristlecones in the lower forests where they mix with other species, but they rarely do so in the pure stands of ancient bristlecones. The final part of the way is resisting or avoiding pest insects and diseases. If you've read the book this far, you may have guessed that our obvious concerns would be mountain pine beetles and white pine blister rust.

Barbara Bentz and colleagues conducted studies in mountains where Great Basin bristlecones stand unscathed among limber pines that died in the mountain pine beetle epidemic. She brought hungry beetles and confined them in "attack boxes" on both limber pines and Great Basin bristlecone pines. Those on limber pines dug right in, but a lot of the ones on bristlecones retreated to the far side of the box. When she forced male and female beetles into cut sections of pines, they bored through both kinds of bark and laid eggs, but in bristlecone pine the larvae failed to thrive. From 110 pairs of parents that went so far as to actually carve an egg gallery, a next generation emerged totaling 53 adults—hardly the sort of numbers that suggest a baby boom—and even those 53 looked malnourished. In all of her sample plots she found zero bristlecones killed by mountain pine beetles. Twenty-two bristlecones were killed by dwarf mistletoe and subsequently dined on by the kinds of bark beetles that go for dead trees, but not by mountain pine beetles.

Chemical analysis told the story: terpenes were eight times more concentrated in bristlecone phloem (the tissue beetles eat) than in limber pine. Bristlecones also had several specific terpenes not found in limber pine. The great density of bristlecone wood further contributes to pest resistance.

All of this is a bit surprising. Theory has it that regions that went thousands of years without a mountain pine beetle outbreak would become full of "naive" pines evolved to devote fewer resources to beetle defense in order to devote resources to other things. That seems to have happened among, say, the northernmost lodgepole pines in Canada. So why did it not happen with bristlecones?

Bentz offers a couple of possible explanations. Both relate to Great Basin bristlecone's extreme longevity. One is that it evolved to be extra pitchy because that worked as a defense against rot fungi—an essential longevity adaptation—and the pitch happens to also defend against beetles. The other is that living thousands of years gives it a longer evolutionary memory, in a way: it was probably subject to beetle attack during and shortly after the ice ages, when it was relatively widespread at lower elevations. For a bristlecone, that isn't so many generations back, not long enough for it to evolve away from that experience. These trees are still reproducing at five thousand years old; they do not senesce. Paleobotanists tell us that the foxtail group has been in the West for forty million years. Bristlecone pine has had many millions of years to evolve its defenses, *and* it received immunity booster shots over the last twenty thousand years to keep it from letting its guard down. The bottom line, in any case, is that we don't have to worry much about pine beetles getting the Methuselah trees.

Their two closest relatives, foxtail pine and Rocky Mountain bristlecone pine, do suffer some pine beetle attacks.

Other scientists have checked in on Great Basin bristlecone to measure how it's doing, with striking mixed results. Growth has sped up since 1950 in the groves at highest elevations, within a few hundred vertical feet of treeline. When you can look at 4,000 years' worth of annual rings, and see that in 3,700 years no shift in the climate ever altered their growth rate as much as the current climate change has, it makes you sit up straight. (There was one period of equally fast growth 3,700 years ago. The study couldn't determine whether that was an artifact of some of the trees being youthful then, which certainly was not the case in the recent fifty years.) The current growth spurt was averaged across hundreds of trees from three mountain ranges. It seems unlikely to result from carbon dioxide enrichment or nitrogen deposition, since it was not shared by the lower-elevation sites in the study. That leaves precipitation or temperature to account for it, and the graph of growth rates in the past century does track impressively well with temperature. Not with precipitation.

The fact that trees at treeline struggle to survive in the cold is obvious: that's why there's a treeline. It makes sense that warming makes it easier for them to grow. By the same token, warming should make it *possible* for them to grow a little bit higher up the slope. (Yes, bristlecones do have room to move up higher on some of these mountains.) And yes, warming has enabled bristlecones to reproduce, including higher up. As late as 1950, no one had seen much evidence of new bristlecones pines getting started anywhere for a long time, maybe five hundred years. It was mystifying and worrisome. Apparently they are only able to regenerate during warmer spells. Now they've gotten one, producing a pulse of regeneration from 1955 to 1978. The lack of new seedlings since 1978 certainly isn't for want of seed production: every ravine I saw in the bristlecone forest was black with deeply piled cones trapped at the ravine bottom.

The ominous results came in recent years when researchers looked specifically for seedlings on slopes uphill from treeline. They found a surprise: limber pine seedlings. Limber pines grow on most of these same mountain ranges, decreasing in number as they approach a top elevation, leaving a narrow treeline belt of pure bristlecone pine. Limber pine's pulse of new seedlings began after bristlecone's and continued longer—from 1963 to 2000. The surprise is that at least locally they leapfrogged up past the pure bristlecone belt and are setting the new, higher treeline. As we saw in chapter 8, limber pine is one of the species that can disperse its seeds extra far because they are big and fatty enough to get themselves cached by birds. Bird caching of the much smaller, winged seeds of bristlecone pines is relatively rare. That's probably one reason why limber pine is moving uphill faster. Bristlecone pine's days of enjoying some habitat without competition from other tree species may be numbered. The Associated Press was quick to conclude that the oldest tree species is "in peril." I still like this species' chances, as western pines go. Though the very oldest specimens are at treeline with no other species to compete with, bristlecone has also held up for hundreds of years in other stands with limber pine competition.

The worst threat bristlecone pines face may be white pine blister rust. It isn't happening yet: no blister rust has been seen on a Great Basin bristlecone pine in its native habitat. Tests in nurseries show that it is susceptible. Rocky Mountain bristlecones in Colorado have already been infected. The other close relative, foxtail pine, proved highly susceptible, at least in the seedling stage, at the nursery in Dorena, Oregon; but very little infection of foxtail pine has yet been seen in the wild. Limber pines are widely infected, share several mountain ranges with bristlecone, and seem likely to bring the pathogen closer.

Some scientists speculate that there's a limit to how much aridity and cold blister rust can take, and that that limit will save the Methuselahs. We can cross our fingers, I guess. Blister rust's southward progress into the highest Sierras has been quite slow, suggesting that the arid, cold conditions there are slowing it down—but won't necessarily stop it. Blister rust eventually infected mountaintops in Arizona that were once said to be too arid and cold. Most plausibly, infection just has to wait a long time for the right combination of conditions; in particular, the right conditions for infecting a currant may take years to join up. In a typical year, high-elevation currants are not yet leafed out in late spring when the spores are released from lower-elevation pines, but eventually spores release and currant leafiness may coincide on a day when there's also fog or drizzle. (Spores need some humidity on the day they infect the currant.)

The tiny spores that carry blister rust from a pine to an alternate host can travel really far in a strong wind—apparently at least a thousand miles, as the disease infected a New Mexico mountain range one thousand miles from the closest known case of it. The heavier type of spores that carry blister rust from the alternate host to a pine travel only a mile or so, we think. There isn't much *Ribes* high in the White Mountains, but there are several kinds of paintbrush and lousewort, the wildflowers that include some species known to host blister rust.

Given the superhuman time scale of Great Basin bristlecone pine re-production and evolution, and the rarity of climate decades when seedlings survive, the idea of saving them via breeding programs is daunting for mere humans. The prospect of a nice crop of youngsters is at best only mildly exciting because it's kind of beside the point: when it comes to bristlecones, it's the ancient individuals we're desperate to save. I'm guardedly hopeful about their prospects. I'd say that some percentage of them will resist blister rust and outlive your great-grandchildren, and that if the bristlecone groves become sparse compared to today, at least they're in an environment where we don't have to worry as much about newcomer species crowding them. There's no tree better at enduring.

Coulter Pine

12 FUTURE FORESTS

What are we to expect, and what influence can we hope to exert?

There will be fire, there will be smoke, there will be pests and disease, there will continue to be more dead trees than we can count.

There will still be a lot of forest, but less than in the twentieth century.

Fires (a normal and typically beneficial element of dry western forests) will in some places catalyze shifts from forest to grassland or brush.

Insect outbreaks will in some places catalyze tree species shifts, such as oaks and incense cedars replacing pines in the Sierra Nevada.

Fires will burn homes in increasing numbers until we get rigorous about making homes and properties much less vulnerable.

Where forests persist, many of them will look sparse; they'll hold fewer trees per acre—and that's a good thing, in much of the West.

Species and genetic types in any given place will shift as they inevitably migrate to fit a changing climate.

There will be invasions of species from other continents.

The climate is not shifting to a new normal that we'll reach this year or next year. It will keep warming beyond 2025 even if we get off fossil fuels tomorrow. All the tools in the forest management toolshed may prove useless beyond a few decades if society continues with business as usual, increasing the greenhouse gases in the atmosphere. So the most critical order of business is converting the world almost entirely to carbon-free energy, while also compensating by actively sequestering some carbon from the air. If that happens, there's hope for the climate to plateau a few decades from now and for some of the actions described here to work on a time scale a pine tree can appreciate.

But to get back to my subject (western North American dry forests, as opposed to the coming energy transition), though the trend toward less forest is inevitable, most people would agree that we want to optimize forest cover where we can.

In some places sustaining forest is a lost cause. We have to do the best job we can of figuring out which places those are, and optimizing their transition to a nonforest state. We can't possibly know for sure. Trees can be amazingly resourceful and durable, showing "phenotypic plasticity." (That's when an individual tree changes its characteristics, such as leaf size or thickness, from one year to the next in response to changes in its environment.) Trees may give us some happy surprises.

The best tools we have for helping our frequent-fire forests—as well as for creating fire-moderating buffers around human populations—are restoration treatments that reduce tree density and favor fire-tolerant species.

These are realities that scientists agree on (while they sometimes disagree on some of the details). It's out in the broader population that I often see unrealistic expectations that obstruct useful action.

Maybe we expect to go back to that cabin by the lake and always have the same view. Or expect to breathe the same clean air around our urban neighborhoods, with the same clear visibility, that we enjoyed in the past. To build our home in the wildland interface out of the materials we like best, or can best afford, and to surround it with whatever landscape plants we happen to like. And then to count on the Forest Service to keep the fires away. To enjoy forest shade on our old favorite hike to the waterfall. To have our community supported by the level of timber harvest that supported it in the 1970s. To treat the environmental laws passed in the 1960s and '70s as untouchable because they are perfect forever. To establish a nature preserve to protect a particular species and then have that protection last, while the ecosystem around it changes with the climate.

It's time to think about what we want from our forests in terms of the options that actually exist. Ask yourself and your neighbors these questions:

Can we accept prescribed fire—a lot more of it than we've seen so far? Those are fires set intentionally by professionals in order to improve the ecosystem and actually to reduce wildfire. Inevitably, every once in a while a prescribed fire will break loose and become a wildfire, like the notorious Cerro Grande fire that burned 235 houses around Los Alamos, New Mexico, in 2000. However, the percentage of prescribed fires that break loose and do serious damage is vanishingly small. In 2012, for example, two million acres were burned in 16,626 prescribed fires, 14 of which escaped to any degree, and none catastrophically. Any rational study of the numbers will conclude that an extensive program of prescribed fires will reduce the total economic and human-life risk from forest fires, while improving forest resilience.

Environmental activist Todd Schulke is painfully familiar with Cerro Grande; yet even though he's never shy about criticizing the Forest Service, he gives high marks to one part of that agency: "Fire managers are getting better and better at doing effective burning even in relatively dense stands."

Inevitably, prescribed fires produce smoke. Again, they prevent more smoke than they produce. Wildfires tend to burn at the peak of summer heat and drought, when they burn hotter and can grow to enormous size. In contrast, most prescribed fires are at cooler times of year. (That's sometimes a drawback ecologically, in failing to mimic historic fire regimes. It's a trade-off managers make for the sake of reducing smoke, fire intensity, and risk—the risk of the fire getting out of control.)

Wildfires are usually thought of as acts of God; prescribed fires are acts of humans. For some people that may pose an emotional obstacle to accepting prescribed fire, despite the favorable net results and despite the fact that the severity of recent wildfires is partly human-caused. To bureaucrats involved in decision-making on prescribed fire, the act of God factor is a serious impediment. The Clean Air Act regulates smoke produced by humans, while giving a pass to wildfire smoke. People whose homes burn down (or their insurers) may sue an agency that started a prescribed fire, but if it was a lightning fire, it's hard to pin blame on humans.

Prescribed fires should be given some slack in light of the overwhelming evidence that in the long run they reduce smoke and reduce risk to property. As Craig Thomas put it, "We can all be grumpy about smoke together. We've gotta pay some dues here to get back to the place we need to be. The only other real choice is to be socked in for weeks at a time while we're burning down a hugely important landscape."

Are we willing to pay for increased use of prescribed fire—and of forest restoration thinning—at several times the current pace? Tom Swetnam makes a vigorous case that it would save money in the long run: "Just here in the Jemez Mountains alone, well over a hundred million dollars in fire suppression costs have been incurred since 2000 (not counting houses and forest/watershed value losses). How much restoration work, thinning, burning could have been done with that? The whole mountain range, multiple times over!"

There are savings in firefighting expenses, in structure fires prevented, floods averted, in higher rates of natural reseeding of the forest. It's easy to see fuel treatments paying for themselves if you include hard-to-quantify things like the amenity value and ecosystem services of the healthier forest. But to the people in green visors, it remains an uncertain case. Clear cost savings have been hard to achieve, so far, because our efforts have been pathetically paltry: as one review article concluded, "in order to save large amounts of money on fire suppression, land management agencies may need to spend large amounts of money on large-scale fuel treatment."

Whether there would be a net improvement to the forest's carbon budget (helping to mitigate climate change) is also debated, based mainly on computer modeling. These studies' conclusions derive quite directly from their assumptions, e.g., what's the probability of the treated area (and of the untreated area it's being compared with) to burn in the future? How severely? How soon? Will they then regenerate as forest, or convert to low-biomass grassland, or to chaparral that will burn again? How soon? and so on. To answer these questions with a controlled experiment would take half a century. Interestingly, scientists who live in regions that have seen big increases in severe fire (sometimes without any forest regrowth in sight) foresee higher levels of future wildfire, and accordingly predict greater carbon-budget benefits from fuel treatments.

But we don't have to wait for a these debates to be settled. Forests are justification enough. **What are our forests worth to us?** Do we care enough to support a budget that could do something to help them?

Will we support planting genetic races northward of their native range? What about species? There's a lot of resistance to that idea. Some people worry because they've heard the many true stories of intentional species translocations gone bad. Almost all of those were from one continent to another. Responding to changing climate by moving north and south within the western cordillera is exactly what our tree species do naturally—what they did over and over again throughout the ice ages—and what they will do in response to warming in the next few centuries. But this anthropogenic warming is fast, and we want to minimize disruptions in ecosystem services. Putting suitable near-native species in place sooner than they can arrive on their own could reduce the plant-community "vacuums" that non-native invasives fill.

I'm not suggesting planting any monocultures, or even picking winners, exactly. Just pick the slate of nominees. Picking one native species as a winner and planting too much of it was one kind of big mistake foresters made in the last century. (Their second-biggest type of mistake, I'd say, after fire suppression.) Heterogeneity is the key! Augment the local genetic offerings with a broader, more southerly menu of them, and then let natural selection pick the winners.

Even without trying to move genes northward, increasing forest heterogeneity in terms of species, ages, and tree spacing would be a big improvement. Either way, replanting has to be better funded than it is today in the

U.S. Forest Service. Species under attack by non-native pests or diseases, like the whitebark pine, critically need well-funded planting programs if they are to survive. The pipeline of seedling stock and planting budgets has to be large and flexible enough to wait and then go to work when the wetter-than-average years come along. In some forests hard hit by insect infestations, heterogeneous regrowth will happen naturally if we forgo salvage logging and leave the forest alone: both the unattacked species and the surviving trees of the attacked species—which probably have superior genes—will thrive.

Can we accept allowing more lightning fires to burn rather than always putting them out? Always putting them out has the perverse effect of increasing fire size and severity. If we put out every fire we *can* put out, then in logic the fires that get big will be the ones we *can't* put out. Those are the ones that burn under extreme fire weather conditions. It's the recipe for megafires and huge high-severity patches. Fires that burn in average summer conditions are more likely to be smaller, beneficial fires; they leave the forest with less fuel in it, making a megafire on that site less likely in the future. Letting fires burn has been the policy for one drainage in Yosemite National Park for forty-seven years, and sure enough, average fire size and severity there has already dropped well below what you see in the fire-suppressed part of the Sierra Nevada. The *total acreage* burned over the forty-seven years is greater than outside the park, but they were "good fires" that left plenty of trees standing. Zion National Park has let fires burn since 1988, and scientists there are able to measure a similar drop in likely fire severity.

Can we stick to our beliefs in prescribed and managed fire?

I have to be honest here. In efforts to sell the idea of beneficial fire, unrealistic promises are sometimes made. Like this one: if society chooses to have more of the right kind of fire, we'll have less of the wrong kind. We expect it to be less, yes, but compared to what? Not less than in the 1980s and '90s. Because of climate change (and partly because of dense, homogeneous forests) we're going to have more smoke and more fire, probably including more high-severity fire, than in the late twentieth century. It's inevitable. We can only have less compared to how bad it will be if we carry on just as we have been. (If we're lucky, maybe it will be less in most years than in bad years like 2012, 2017, or 2018.) We'll never be able to prove it; we'll never know for sure how bad it would have been if we had chosen otherwise. It may be tough for us to stay the course while other people are yelling about how much worse they think the wildfires got after we stepped up our program of fuel treatments. Fires may in fact trend worse from year to year, over some periods in some areas.

But we have the evidence. We have a very strong consensus among fire ecologists, based on myriad studies, that thinning and burning are the best ways available to us of limiting the severity of future wildfires, and of averting the worst future episodes of hazardous smoke. Thinning and burning may not limit the number of days with some smoke, or the number of acres burned. The object is to burn more acres with milder fires so that fewer of them burn in lung-damaging, forest-destroying fires.

Underlying all the preceding questions is this question: can we trust government agencies to direct these interventions?

There's no one else who can do it on a large scale.

Can local jurisdictions take over more costs of fighting fires? We have a mismatch where one governmental entity pays and a different one regulates. State and local governments have too little motivation to enact and enforce the building codes that could save homes from fire, as long as the federal government is picking up the very large tab not only for firefighting but to a large extent also for rebuilding. Sometimes the replacement homes are no more fire-safe than the ones that burned down. The federal government pays for firefighting but has no power to enact building codes.

Insurance companies would seem to have the motivation to get people to make their homes safer, but surprisingly little of that has happened so far. Insurers' losses to wildfire, though considerable, are actually small potatoes, relatively speaking—at least outside of California. The bigger potatoes are hurricanes, floods, earthquakes, even hailstorms. In Colorado, for example, the shocking suburban fires in 2012 and 2013 come in at numbers four and five in the rankings of Colorado's costly insurance events; the eight others in the top ten were all hailstorms. Headwaters Economics, a Montana think tank, analyzed the issue and determined that insurance reform is not one of the more promising paths toward fire-safe communities. In my view, it would help if states would at least clarify that insurance companies are allowed to set lower premiums for properties that meet high fire-resistance standards.

Can those of us who own land in and near the forest take responsibility for making homes there as fireproof as possible, and our acreage a low-fire-severity habitat type?

The measures needed to make a house near a dry forest fire-safe fall into three categories: construction and materials, clearing vegetation, and maintenance. Building codes can only affect the first of the three. Maintenance means things like keeping the roof and gutters clear of pine needles—which is crucial because the number one way houses catch wildfire is from flaming embers landing on them. Some fires subject houses to veritable blizzards of embers, and airflow through the house draws embers inside, where they set the house afire. New vent designs that can screen embers out are available, ranging from simply smaller-mesh screens to louvers that swell in the heat to close the vent automatically.

While strong building codes are essential, the main way to affect maintenance and vegetation is through community education, spreading the word, social pressure. It's only fair, because your neighbor's house or patch of forest can bring the fire to your house. It's a little like vaccination: in order for it to work well, you need a high percentage of the community participating.

Mayor Walt Cobb of Williams Lake, B.C., told the *Vancouver Sun* that "removing trees and other flammable brush can cost property owners money and require permits or licences, so there is little motivation to use FireSmart initiatives beyond self-preservation." If self-preservation isn't a strong enough motivator, what in the world would be?

Of course, what Mayor Cobb was hinting at was bigger government subsidies. In the United States, the National Institute of Building Sciences says that subsidies are an excellent investment: they estimate a three-to-one or four-to-one benefit-to-cost ratio for disaster recovery costs avoided via wildfire mitigation. Programs called FireSmart in Canada and Firewise or CPAW (Community Planning Assistance for Wildfire, planningforwildfire .org) in the United States provide support and information in great detail.

Effective structural improvements are also possible outside the houses themselves: bury power lines so that wind events can't set them to sparking, or so that water pumping can continue during fires; store water in rooftop tanks to spray on houses during fires; upgrade water systems with fire hydrants and plenty of water capacity for firefighters; find less flammable plants for landscaping. "Green firebreaks" are being studied and tested in Australia and New Zealand but have barely been looked at for western North America. At this point only a few scraps of advice are available and they're pretty obvious: cacti are barely at all flammable. Aspens and irrigated fruit trees are good. Junipers, cypresses, palms, and eucalypts are highly flammable and should be avoided, or isolated well away from structures. Fences, decks, and lawn furniture can also be fire hazards.

Envision fire-resistant communities like this: compact patches of houses with a modest buffer zone separating them from surrounding forests regularly maintained with prescribed fire. A grassy park can serve as buffer in areas that have enough water to irrigate it, or a xeriscaped park in very dry regions. Isn't that far more appealing than entire viewsheds with few trees because the whole area is divided into five-to-twenty-acre homesteads that all have to be drastically thinned to protect the homes from fire? Or that have few trees left after a fire actually comes through?

As I think about western forests of the twenty-first century, I place myself back among those gorgeous gnarly bristlecone pine matriarchs looking out on the world thousands of feet beneath their perches on the White Mountains, as they've done for more than five thousand years.

Five *thousand* years.

How am I to comprehend such age? I'm just a writer. My tools are words and letters whose first prototypes were scratched during the same misty millennium when those pines germinated. Individual Bronze Age trees live on, their lives spanning all of human literacy as well as the hockey stick of human-caused warming. Countless living trees in the West, of many species, predate Columbus. And suffer predation from Columbus's heirs.

The twenty-first century is a challenging time for forest trees, and we need them more than ever. For refuge, for their beauty, for clean water, for their ecosystems, yes, all of the above. They're the source of our oxygen, the sink for our carbon. Literally, we can't even breathe without them.

AFTERWORD

The summer of 2019 was one of the worst on record for fires—globally. The total area burned was staggering. It was mostly far away, in the boreal North and in the tropics. As summer shifted to the southern hemisphere, Australia's fire season exploded early, following record drought and heat. In California, the fall season inflicted a rash of ferocious human-caused, wind-driven fires and smoke—the kind that increasingly typifies autumn in California. Those fires mainly burned brush and buildings rather than forests. If we look strictly at the West's pine forests, they got quite a reprieve in 2019, but the surfeit of fire elsewhere reaffirms (if any further confirmation were needed) that an upsurge in fire is an inescapable element of the climate crisis.

In September 2019, I turned off the freeway into North Redding, California, near the site of the Carr fire tornado of 2018. Right away, I felt disturbed, like something was wrong, or perhaps that something was right where it should be wrong. What's going on here? Everything looks disturbingly normal: franchises doing a tidy business, traffic gliding smoothly, residents jogging, shopping, walking dogs, everything intact and clean, the plants green. How can this be, so close to the incineration zone that I know lies in wait just over the next hill?

It came to me that this serene vision of untouched, untouchable suburbia was exactly how the people who lost their homes saw their own neighborhoods up until the day, or even the hour, before an ungodly pillar of fire struck them down.

I turned onto Quartz Hill Road and soon entered the devastation. Walking past a barricade, I came to a dead ghost pine with a thirty-foot strip of shiny roof flashing entangled in its high limbs. Beyond that point, trees were whittled down to charcoal spindles. Underfoot there was nothing but bare mineral soil, scattered plants withering in late summer. I came upon a circle of ceramic saints and Savior; a memorial?

There were still rusted skeletons of machines, but overall, plenty of cleanup had been done over the thirteen months; many replacement homes were complete, many others underway. All had nonflammable roofs and siding and sparse plantings. A few patches of green sod stood out, like golf greens on the desert skirts of Vegas. One owner had moved a motor home onto their old house site and planted a brave row of six coast redwood saplings along the property line. They looked desperate for a little shade, a little coastal fog.

Two hours south I found Paradise, site of the Camp fire. I followed my nose: an extra-toasty ponderosa pine fragrance filled the air for miles. Reconstruction here was much proclaimed on signs, but relatively little accomplished compared to Redding, and I don't think that was primarily because the Camp fire was more recent by three months than the Carr fire. For one thing, this was a less prosperous demographic, the homes less well insured. For another, Redding's refugees had a big town to rebuild in, whereas in Paradise most of the businesses were gone, along with 90 percent of homes, and it was hard to count on them returning at any predictable time. With no town to move back to, many were finding it hard to make the decision to move back.

Obliterated homes, mile after mile, came as a shock, even though I expected them. The Mamma Celeste's Gastropub and Pizzeria sign stood proud and shiny, but failed to get my mouth watering; no pizza ovens sizzled behind it, just a razed parking lot.

The unexpected was the overstory—the many trees overhead, still green. They made it easy to see why most residents used to feel that the town lived up to its name. My flimsy estimate from driving around town was that canopy tree mortality was in the 50 percent ballpark. In ecological fire severity terms, that means this wasn't even a high-severity forest fire. Assuredly it was severe where it approached the town, raging and racing across six miles of primarily brush and small trees grown up in the decade since a previous fire. But once it passed the limits of that 2008 burn and entered Paradise, the trees were older, more fire resistant, and the fire became more of a house-to-house ember storm.

While many people talk about how homes and lives could have been saved if the power lines that ignited the fire had been maintained better—and that is true—a truly hard-core campaign to harden structures against fire would also have made a huge difference. There's still time for such campaigns in the hundreds of other Western towns that rate just as vulnerable as Paradise ever was.

Californians are settling into the grim reality that a great many of their homes face a fire threat that will be difficult or prohibitively expensive to insure against. Meanwhile, they suffer a wasteful and inequitable program of power blackouts attempting to reduce the fire threat. It isn't succeeding yet.

This year brought a bit of consolation in a broad study asking which factors most affect a California home's chances of surviving a fire. It turns out that certain house retrofits can help a lot, and not just the ones that cost thousands, like new roofs and siding. The greatest benefit-to-cost ratio comes from installing eighth-inch mesh screens in all vents. Scrupulous removal of flammable litter, debris, or dead branches during fire season doesn't take much money, just a lot of time and energy. Ornamental shrubs near the house are not necessarily ruled out (with some exceptions like juniper); they just need to be kept well watered. Double or triple window panes, preferably of glass rated for high heat, also prove critical to house survival. Installing those is expensive, but they gradually pay for themselves by reducing home heating and cooling costs. Their energy efficiency also helps with greenhouse gas emissions, making a strong case for subsidizing them.

The study did not look at thinning and burning in the surrounding forests as a factor for saving houses. Those fuel reduction treatments can help, but they probably help more in other states where there's less chaparral and far fewer wind-driven, human-caused autumn fires. California's autumn fires are an almost completely separate issue—brush fires that turn into urban fires—whose solutions lie in such areas as community design, structure retrofits, infrastructure maintenance, and faster detection when the fires are still tiny. As for forest fuels reduction treatments, saving houses is one potential benefit; but their primary objective is to help our pine and dry-mixed-conifer forests survive into the future.

The disastrous 2017 and 2018 fire seasons lit a fire under the California legislature, which passed serious funding for a broad menu of responses to the fire threat, specifically including thinning and burning. By September, Forest Service ecologist Hugh Safford was able to say, "I can't remember a summer when there's been so much prescribed fire. We rarely burn in July and August, but this summer . . . we burned sixty thousand acres. A minuscule amount compared to what we should be burning, but three times more than what we typically have done in a year." He also detected "a keen interest in taking advantage of natural burns." On top of the new funding, prescribed burning benefitted from a quiet wildfire season across the West, which left plenty of fire crews available.

Silviculturist Dana Walsh emphasized the shortage of trained burn bosses and burning crews as a key impediment to prescribed burning. "We've got four districts" in the Eldorado National Forest, she told me, "and all of them have tens of thousands of acres on the books that need to be burned."

Craig Thomas, the Sierra Nevada forest activist, exuberantly told me about the Caples fire in one of the Sierra's most beautiful valleys. He had taken part in years of planning and preparation for it. Paradoxically, it was especially gratifying because it went "wrong" during the October red flag warnings that brought terrible fires to Sonoma County. Initially a prescribed fire, it was reclassified as a full suppression fire when the east wind developed and it overstepped its planned boundary. Despite working during power cuts and high winds, the Forest Service managers successfully contained it—yes, at a larger footprint than originally prescribed, but still with good ecological effects throughout. Thomas sees the unplanned portion as a 2,500-acre bonus. Perhaps the bigger bonus was that it demonstrated fire managers' capabilities under duress.

The New Mexico government did equally well on a smaller stage. They enacted the Forest and Watershed Restoration Act and funded it with an annual sum set to automatically repeat every year. Projects were hitting the ground already in 2019.

Oregon chalked up a fortuitous success at Granite Gulch, a lightning fire in wilderness in the Wallowa Mountains. The Forest Service decided to let it burn as a managed wildfire for seven weeks, and they did manage it, a lot, closely monitoring it with computer models and nudging it here and there with helicopters dropping lake water. By mid-September, it burned 5,500 acres, with broadly good effects. It reached mountaintops where whitebark pines mix with subalpine fir, providing a test of whether fire can sometimes help whitebark to hold on by weeding out some of the firs. Smaller lightning fires in ponderosa pine in south-central Oregon were also managed successfully.

British Columbia, like California, suffered its worst one-two punch of fire years in 2017 and 2018. Unlike California, B.C. emerged gun-shy. "There's very little appetite for prescribed fire," fire ecologist Daniel Perrakis told me. The Canadian forest scientist community knows that fire is needed in the dry interior forests, but this awareness has made few inroads into the body politic. B.C. does have a revamped program directing subsidies into thinning, biomass utilization, and other actions that may have beneficial effects on carbon management and forest health. These are clearly focused on keeping the timber industry profitable as well.

In most states, 2019 brought no dramatic increase in funding for prescribed fire or other restoration treatments. Several, however—like Colorado, Montana, Washington, and Oregon—created or revamped programs hoping to get more work done through partnerships between various entities like counties, tribes, businesses, fish and wildlife services, and other state and federal agencies.

Overall, it seems that dramatic increases in funding await leadership at the national level. The U.S. farm bill passed in 2018 is supposed to start freeing up money for restoration in 2020 by protecting the overall Forest Service budget from getting raided for firefighting, but there was no new money in 2019. President Trump tweeted that he blames California's fires on insufficient forest management by the state. His notions that we need to "do burns" or to "rake" forests are not far off the mark, but he betrayed a staggering ignorance as to who is supposed to do this burning and raking, and it seemed no one around him bothered to inform him over the year that elapsed between these tweets: it's the U.S. government, not the state, that manages most forestland in California and other Western states. Trump's own administration has been cutting the Forest Service budget, throttling its ability to do burns.

The year's biggest media commotion in the sphere of forests and climate change may have been sparked by a study claiming that the world's "most effective climate solution" would be to plant a trillion trees. The authors had mounted a massive computerized analysis applying machine learning to satellite imagery of global vegetative cover. I'll call it the Bastin paper, after its first-named author, Jean-François Bastin. After excluding cities and productive farm land, they found patches of ground all over the earth that currently support fewer trees than they may be able to support, based on comparison with climatically similar protected areas nearby. They calculated that planting a trillion trees or so in those patches would bring them up to their potential "natural tree cover," and that if those seedlings were able to grow to maturity they would sequester 205 gigatonnes of carbon, around two-thirds of the gigatonnes that humanity has released, they say, by burning fossil fuels.

The authors are members of a Zurich lab funded by a Dutch foundation and led by a U.K. citizen born in Namibia and raised in Wales. The lab employs public-relations staff, which succeeded in inspiring an immediate global paroxysm of media euphoria. That may have sharpened the bitterness in the reaction of the scientific community over the next few weeks.

Just weeks earlier, in a paper signed by a dozen forest ecologists I talked to for this book, I had read that "plans to increase planting densities to sequester more carbon will likely result in elevated bark beetle– and wildfire-related carbon losses, rather than gains." No surprise there, but certainly a strong dissonance with the trillion-trees proposal, at least as far as it applies to the West's pine lands. So I contacted several coauthors of that sentence for their reaction to Bastin.

First, a couple of emphatic affirmations: Yes! Forests are a major, major carbon sink. They're the biggest category of ongoing carbon sinks on land (though much smaller than those at sea). Yes! Greater support for tree planting in our region is direly needed—especially for planting within the footprints of large severe fires that leave too few seed trees.

Craig Allen, the Santa Fe seer of global forest mortality, replied that he appreciates the level of detail that went into Bastin's "important work: previous work on this topic globally has been much coarser and less nuanced." Hugh Safford told me, "It's neat to know that at least some models suggest you could make a global difference planting trees, as long as you can keep the trees alive."

But when they look specifically at our dry forest region, our ecologists think Bastin's proposal goes in the wrong direction. After all, they've been consistently using the word "restoration" to mean *reducing* tree density

within the frequent-fire forests, in order to restore their presettlement condition. Now here's Bastin advocating "tree restoration," meaning *increasing*
the tree density—which, in our region and quite possibly in other analogous
regions will likely emit more carbon than it stores, our scientists believe, because of our density-correlated fire and insect threats. Bastin wrote one little
nod in our direction, acknowledging that in some regions the protected areas (what they used as models of natural density) display *unnaturally* dense
forest "because of low fire frequency," but their analysis did not adjust for
that factor.

Around the globe, harsh reactions sprouted like fireweed, with two
main themes: that the group gets the numbers wrong and that their grand
statements are dangerously sloppy—phrases like "the most effective solution
at our disposal to mitigate climate change."

The danger lies in the way any kind of optimistic science on climate
mitigation—especially if it doesn't involve fossil fuels—tends to morph as
it bounces around the internet. Within minutes, people are saying that we
don't have to curtail our emissions anytime soon; we can just plant trees for
now. The Bastin group did not explicitly rule out that interpretation in their
paper or their first round of PR, but soon they saw their error. Team leader
Thomas Crowther told a reporter that "we certainly did our communications a little bit wrong, and we've learned from that. I want to be extremely
clear that cutting greenhouse gas emissions is absolutely essential if we are
going to have any chance to stop climate change."

For an example of sloppy numbers, a scientist from Congo complained
that his city, Kinshasa, with its twelve million souls, appears on Bastin's map
as a nonurban area ripe for reforestation. That letter claimed to tally 2.5 billion humans occupying areas that Bastin's map failed to identify as urban.
When Craig Allen zoomed in on familiar New Mexico areas, overly optimistic algorithms appeared to be in play, finding restoration potential where he
was pretty sure it's too hot and dry for new seedlings to thrive.

A look at the broad assumptions underlying the study raises another
troubling issue. Their figure of 205 gigatonnes of carbon is based on the
trillion trees *at maturity*, which would be many decades away. They then

looked into the effects of ongoing climate change on the world's trees over the course of those same decades and concluded that, on net, tree growth will slow, and the area capable of growing trees will shrink. They nevertheless chose to assume current climate, rather than expected climate, in their analysis. It's an understandable decision, since current climate is a more solidly known quantity. But it means that their conclusions stand on a basis that's not just hypothetical, but impossible: a stable climate and stable land use over future decades. And they chose not to calculate or even speculate as to a realistic-climate adjustment of their 205-gigatonne sum.

In my view, all their grand statements about restoration potential should carry a warning that they are intentionally based on unrealistic assumptions, rather than on our actual near-future world in which any gains from planting trees will be eaten into by ongoing decreases in the world's capacity to grow trees—both any newly planted trees and today's intact forests.

Nevertheless, bottom line, the simplest message from Bastin's study is one I enthusiastically endorse: plant trees.

In planting trees, aim to perpetuate an ecosystem, not a plantation. (I would say "perpetuate the local ecosystem," except that it's well worth borrowing from a somewhat warmer ecozone.) Many scientists have found that native forests do a better job of sequestering carbon than single-species plantations, and letting older forests keep growing may do at least as much over the near term as planting new forests.

And in planting trees within the dry forests of western North America, do not try to increase forest density. For the existing forests to persevere, many of them need to lose some of their trees and become "pine openings" again, like the ones that inspired a pioneer to write that they "never saw anything more beautiful," and another to write "the atmosphere was loaded with balm."

In September 2019, I wandered western pine lands sniffing for that balm on the air. I went up to big pines and put my nose into bark crevices. I carried a little bottle of vanilla extract for comparison: does ponderosa pine or Jeffrey pine share more aromatic notes with vanilla? In other groves, I didn't need to get close; the balm pervaded, bathing me. I kicked pine cones off to the side and fluffed my sleeping bag onto the bed of dry needles. I could easily believe the forest fragrance was a balm for my health. The Japanese call this *shinrin-yoku*, "forest bathing," and in scientific studies they identify specific benefits ranging from lowering blood pressure to shrinking tumors, at least in mice. (The abundant terpene alpha-pinene has no direct effect on tumor cells in a lab, but when smelled, it starts a chain of effects via the nervous system.)

Since the Industrial Revolution and possibly longer, we have loaded the atmosphere with less-fragrant compounds—nontoxic ones, mostly, but we've thrown them out of balance on an almost inconceivably grand scale. We've given the entire earth system a hard kick in the ribs. Ecosystems will survive in some form, but forests will migrate and change, change in ways we may not like. For us to continue enjoying forests in the ways we count on, we'll have to change—not only by unloading the atmosphere, but also by working with our forests, in our forests, to keep some semblance of those familiar forests as a part of our developing future.

Acknowledgments

The idea of this book germinated over dinner with my beloved friend Gary Braasch, an inspired nature photographer who rededicated his career to documenting the growing climate crisis. Together we developed a book concept for a version of *Trees In Trouble* full of images by Gary. He introduced me to his old friend Jerry Franklin, who joined us under the pines on the banks of the Metolius. The book took a different form after Gary's untimely death in the waters of the Great Barrier Reef. Thank you, Gary, for the beginnings. I long to see the images you would have produced.

At critical junctures, friends supported the further development of the book: Thank you, John Daniel, Sarah Stephan, Jenny Wapner, and K. D. Ray VI. And thank you Sabrina my wolf, for so many years of support of so many kinds.

Tremendous thanks to Matt Strieby, a tree expert in his own right, for the gorgeous pines throughout the book.

It has been a sublime privilege to see forests with the guidance of leading ecologists and experts who showed me around their territories or their labs:

Jerry Franklin, Diana Tomback, Hugh Safford, Andrew Merschel, Richard Sniezko, Craig Allen, Ellis Margolis, Tom Swetnam, Greg Greene, Greg O'Neill, Diana Six, Colin Maher, Tova Spector, Keala Hagmann, Rick Brown, Alina Cansler, Andrew Larson, James Benson, Katie Sauerbrey, Craig Bienz, Russ Parsons, Carl Seielstad, Adam Watts, Kellen Nelson, and Aaron Ramirez.

Profound thanks go to most of you on the above list for also reviewing parts of the text and providing valuable corrections and suggestions. Now, that was just the ones who talked to me in person. Others reviewed text after generously giving of their time for phone interviews: Sally Aitken, Barbara Bentz, Craig Clements, Lisa Floyd-Hanna, Emily Heyerdahl, Garrett Meigs, Connie Millar, Nate McDowell, Malcolm North, Daniel Perrakis, David L. Peterson, Susan Prichard, Todd Schulke, Nate Stephenson, Scott Stephens, and Dana Walsh.

Additional generous phone interviewees were Lori Daniels, Kim Taylor Davis, Jim Flanagan, Jason Forthofer, Dan Gavin, Brian Harvey, Sharon Hood, Aaron Liston, Sean Parks, Camille Stevens-Rumann, Anna Talucci, Alan Tepley, Craig Thomas, Diane Vosick, and Jessica Wright.

And still more scientists provided answers, info, and encouragement via email:

John Abatzoglou, Steve Acker, Pico Alt, Bill Baker, Phil Cannon, Tony Caprio, Derek Churchill, Janice Coen, Jonathan Coop, Lisa Doner, Paula Fornwalt, Han-Sup Han, Winslow Hansen, Matthew Hurteau, Ann Kennedy, Zeke Lunder, Martin MacKenzie, Jennifer Marlon, John Marshall, Kerry Metlen, Jonathan Nesmith, Alistair Smith, Ron Wakimoto, Tongli Wang, and Park Williams. Many others have done important work in this field and appear to be "missing" from this book: for no good reason, they just didn't end up in my conversations. David Breshears, Paul Hessburg, or Don McKenzie come to mind.

I first heard of publishing legend Jack Shoemaker decades ago, from Lew Welch. Little did I dream that I would one day join Jack's illustrious list of authors. Thank you, Jack, for resonating with my sense of the need for this book. Thanks to Jenny Alton, Jenn Abel Kovitz, Jordan Koluch, Megan Gendell, Yuki Tominaga, and the rest of the Counterpoint/Catapult team for meticulous attention to bringing out the best in my efforts.

Notes

Introduction

3 *Lew Welch and Gary Snyder, as told by Bill Yake.* Bill Yake, *This Old Riddle: Cormorants and Rain* (Astoria, OR: Radiolarian, 2004), 24.

6 *there are some ominous signs.* Kimberley T. Davis, et al., "Microclimatic Buffering in Forests of the Future: The Role of Local Water Balance," *Ecography* 42, no. 1 (2019): 1–11.

Chapter 1: A Loaded Atmosphere

10 *The plates' width is a clue to the tree's age.* Robert Van Pelt, *Identifying Old Trees and Forests in Eastern Washington* (Olympia: WA State Dept. of Natural Resources, 2007), 79–90.

11 *Our drive of the forenoon of [September 8, 1853] was still among the pine openings.* H. K. Hines and R. Ketcham, quoted in Harold J. Peters, *Seven Months to Oregon: 1753* (Tooele, UT: Patrice Press, 2008), 270–71.

16 *The 1859 discovery of the Comstock Lode.* Thomas J. Straka, "Timber for the Comstock," *Forest History Today* (Fall/Winter 2007): 5.

17 *"the tomb of the forest of the Sierra."* William Wright, *The Big Bonanza* (1876; reprint, New York: Alfred A. Knopf, 1947), 174.

21 *Scott Stephens gave a conference talk that crescendoed.* Scott Stephens, "Reform Fire and Forest Policy to Emphasize Resilient Forests Long Term," presentation at AFE Fire Congress, filmed 2017, Orlando, FL, video, 20:26, mediasite .video.ufl.edu/Mediasite/Play/6b121a2f5cc944a181d380758cc4cec01d.

24 *the pine forest is likely to give way to brush or . . . to a new forest with far fewer pines.* C. A. Lauvaux, C. N. Skinner, and A. H. Taylor, "High Severity Fire and Mixed Conifer Forest-Chaparral Dynamics in the Southern Cascade Range, USA," *Forest Ecology and Management* 363 (2016): 74–85.

24 *the future forest needs to be far more sparse, clumpy, patchy, and diverse.* Malcolm P. North et al., "Tamm Review: Reforestation for Resilience in Dry Western US Forests," *Forest Ecology and Management* 432 (2019): 209–24.

25 *Of all the U.S. fires since 2000 where more than $20 million was spent on firefighting.* Jerry T. Williams and M. H. Panunto, "Assessing High-Cost Wildfires in Relation to the Natural Distribution of Ponderosa Pine in the 11 Western States (2000–2017)," *Wildfire* 27, no. 3 (2018): 22–31.

25 *4 to 8 percent of fire area within the mixed-conifer forest burned at stand-replacing severity during presettlement times.* C. Mallek et al., "Modern Departures in Fire Severity and Area Vary by Forest Type, Sierra Nevada and Southern Cascades, California, USA," *Ecosphere* 4, no. 12 (2013): 1–28. Also Jay D. Miller and Hugh D. Safford, "Corroborating Evidence of a Pre-Euro-American Low- to Moderate-Severity Fire Regime in Yellow Pine-Mixed Conifer Forests of the Sierra Nevada, California, USA," *Fire Ecology* 13, no. 1 (April 1, 2017): 58–90. For OR and WA: Ryan D. Haugo et al., "The Missing Fire: Quantifying Human Exclusion of Wildfire in Pacific Northwest Forests, USA," *Ecosphere* 10, no. 4 (2019): e02702.

Chapter 2: Inferno

29 *In their time-lapse movies (which you can watch online) the monstrous smoking convection column.* Michael Zeiler, "June 26, 2011 Las Conchas Fire near Los Alamos," published June 26, 2011, video, 2:29, www.youtube.com /watch?v=mju9oYwI36c.

30 *This one, astonishingly, blew up into a firestorm at 2:00 a.m.* Kyle Dickman, "How a Wildfire Kicked Up a 45,000-Foot Column of Flames," *Popular Science,* June 21, 2017.

32 *A paper he titled in a relatively muted tone has been cited 4,202 times.* (Last counted on November 14, 2019.) Craig D. Allen et al., "A Global Overview of Drought and Heat-Induced Tree Mortality Reveals Emerging Climate

Change Risks for Forests," *Forest Ecology and Management* 259, no. 4 (2010): 660–84.

34 *A California study estimates that thinning forests to presettlement density.* J. W. Roche, M. L. Goulden, and R. C. Bales, "Estimating Evapotranspiration Change Due to Forest Treatment and Fire at the Basin Scale in the Sierra Nevada, California," *Ecohydrology* 11, no. 7 (2018): e1978.

34 *fires converted these sites back to soggy meadows.* Gabrielle Boisramé et al., "Managed Wildfire Effects on Forest Resilience and Water in the Sierra Nevada," *Ecosystems* 20, no. 4 (2017): 717–32.

35 *when piñon-juniper woodland was sharply thinned by bark beetles.* L. Morillas et al., "Tree Mortality Decreases Water Availability and Ecosystem Resilience to Drought in Piñon-Juniper Woodlands in the Southwestern U.S.," *Journal of Geophysical Research: Biogeosciences* 122, no. 12 (2017): 3343–61.

35 *it attracted the attention of some very smart physicists.* Kyle Dickman, "How a Wildfire Kicked Up a 45,000-Foot Column of Flames," *Popular Science*, June 21, 2017.

36 *Flying above the Pioneer fire, they measured the updraft.* SJSU FireWeatherLab (@Fireweatherlab), April 30, 2018, Twitter. Also Douglas Fox, "Inside the Firestorm: New Technology Allows Scientists to See the Forces Behind the Flames," *High Country News*, April 3, 2017.

36 *Now, with the aid of satellites, we see that pyrocumulonimbus events are doing that.* Michael Fromm et al., "The Untold Story of Pyrocumulonimbus," *Bulletin of the American Meteorological Society* 91, no. 9 (2010): 1193–210.

37 *a review of what's known about vorticity in wildfires.* Jason M. Forthofer and Scott L. Goodrick, "Review of Vortices in Wildland Fire," *Journal of Combustion* (2011): 1–14, doi.org/10.1155/2011/984363.

38 *An analysis of how this tornado formed sees several factors joining forces.* N. P. Lareau, N. J. Nauslar, and J. T. Abatzoglou, "The Carr Fire Vortex: A Case of Pyrotornadogenesis?," *Geophysical Research Letters* 45, no. 23 (2018): 13107–15, doi.org/10.1029/2018GL080667.

39 *Five separate pyrocumulonimbus clouds developed.* David A. Peterson et al., "Wildfire-Driven Thunderstorms Cause a Volcano-Like Stratospheric Injection of Smoke," *npj Climate and Atmospheric Science* 1 (2018): 30, doi.org/10.1038/s41612-018-0039-3.

41 *California found that 41 percent of random plots had no seedlings within five years.* K. R. Welch, H. D. Safford, and T. P. Young, "Predicting Conifer Establishment Post Wildfire in Mixed Conifer Forests of the North Amer-

ican Mediterranean-Climate Zone," *Ecosphere* 7, no. 12 (2016): e01609, doi .org/10.1002/ecs2.1609.

41 *A study of eight fires spread across Arizona was even more ominous.* C. Haffey et al., "Limits to Ponderosa Pine Regeneration Following Large High-Severity Forest Fires in the United States Southwest," *Fire Ecology* 14, no. 1 (2018): 143–63.

41 *The broadest recent study covers fifty-two fires in the northern Rocky Mountains.* C. S. Stevens-Rumann et al., "Evidence for Declining Forest Resilience to Wildfires Under Climate Change," *Ecology Letters* 21, no. 2 (2018): 243–52.

42 *Most sites in the study are projected to be too hot.* Kerry B. Kemp et al. "Climate Will Increasingly Determine Post-Fire Tree Regeneration Success in Low-elevation Forests, Northern Rockies, USA," *Ecosphere* 10, no. 1 (2019): e02568. 10.1002/ecs2.2568. Similar results are seen in lodgepole pine and Douglas-fir at Yellowstone: Winslow D. Hansen and Monica G. Turner, "Origins of Abrupt Change? Postfire Subalpine Conifer Regeneration Declines Nonlinearly With Warming and Drying," *Ecological Monographs* 89, no. 1 (2019): e01340. Further refinement in climate thresholds for ponderosa and Douglas-fir came in Kimberly T. Davis et al., "Wildfires and Climate Change Push Low-Elevation Forests Across a Critical Climate Threshold for Tree Regeneration," *Proceedings of the National Academy of Sciences*, 116, no. 13 (2019): 6193–98, doi/10.1073/pnas.1815107116.

42 *Fire can serve as the catalyst for climate-driven shifts in forests.* S. D. Crausbay et al., "Fire Catalyzed Rapid Ecological Change in Lowland Coniferous Forests of the Pacific Northwest Over the Past 14,000 Years," *Ecology* 98, no. 9 (2017): 2356–69. Also many others, including C. Whitlock et al., "Holocene Vegetation, Fire and Climate History of the Sawtooth Range, Central Idaho, USA," *Quaternary Research* 75, no. 1 (2011): 114–24.

42 *many will have fewer tree species than before.* S. Liang, M. D. Hurteau, and A. L. Westerling, "Response of Sierra Nevada Forests to Projected Climate–Wildfire Interactions," *Global Change Biology* 23, no. 5 (2017): 2016–30. Also Alexandra K. Urza and Jason S. Sibold, "Climate and Seed Availability Initiate Alternate Post-Fire Trajectories in a Lower Subalpine Forest," *Journal of Vegetation Science* 28, no. 1 (2017): 43–56.

43 *Gambel oak and New Mexico locust in Arizona and New Mexico.* C. H. Guiterman et al., "Long-Term Persistence and Fire Resilience of Oak Shrubfields in Dry Conifer Forests of Northern New Mexico," *Ecosystems* 21, no. 5 (2018): 943–59.

43 *If the hot zone two inches from the ground exceeds its tolerance limit, the seedling*

dies. P. F. Kolb and R. Robberecht, "High Temperature and Drought Stress Effects on Survival of *Pinus Ponderosa* Seedlings," *Tree Physiology* 16, no. 8 (August 1996): 665–72.

44 *Locally, Oregon's Biscuit fire turned the soil surface into sharp-edged parking lot gravel.* B. T. Bormann et al., "Intense Forest Wildfire Sharply Reduces Mineral Soil C and N: The First Direct Evidence," *Canadian Journal of Forest Research* 38, no. 11 (2008): 2771–83.

44 *On some Colorado mountains it's taking over from conifers after beetle epidemics.* B. Buma and C. A. Wessman, "Differential Species Responses to Compounded Perturbations and Implications for Landscape Heterogeneity and Resilience," *Forest Ecology and Management* 266 (2012): 25–33.

44 *at Yellowstone, it was able to replace conifers in some severe fire patches.* William H. Romme and M. G. Turner, "Ecological Implications of Climate Change in Yellowstone: Moving into Uncharted Territory?," *Yellowstone Science* 23, no. 1 (2015): 6–12.

44 *in Wisconsin it's growing faster than ever thanks to CO_2 enrichment of the air.* Christopher T. Cole et al., "Rising Concentrations of Atmospheric CO_2 Have Increased Growth in Natural Stands of Quaking Aspen (*Populus tremuloides*)," *Global Change Biology* 16, no. 8 (2010): 2186–97, doi.org/10.111 1/j.1365-2486.2009.02103.

44 *Overall, aspen is not one of the more vulnerable tree species.* D. Kulakowski, M. W. Kaye, and D. M. Kashian, "Long-Term Aspen Cover Change in the Western US," *Forest Ecology and Management* 299 (2013): 52–59.

44 *big [aspen] increases are projected in central Alaska.* Zelalem A. Mekonnen et al. "Expansion of High-Latitude Deciduous Forests Driven by Interactions Between Climate Warming and Fire," *Nature Plants* 5.9 (2019): 952–958.

45 *photos from 1935 show brush fields whose outlines didn't budge.* C. H. Guiterman et al., "Long-Term Persistence and Fire Resilience of Oak Shrubfields in Dry Conifer Forests of Northern New Mexico," *Ecosystems* 21, no. 5 (2018): 943–59.

45 *A key change we're seeing now . . . is the size of the high-severity patches.* Brandon M. Collins et al., "Alternative Characterization of Forest Fire Regimes: Incorporating Spatial Patterns," *Landscape Ecology* 32, no. 8 (2017): 1543–52.

46 *Jeffrey pine has been replanted in three different years, and hardly any seedlings survived.* C. R. Restaino and H. D. Safford, "Fire and Climate Change," in *Fire in California's Ecosystems*, 2nd ed., ed. J. Van Wagtendonk et al. (Oakland: University of California Press, 2018), 493–506.

46 *These studies found that post-fire landscapes literally often waited three or four*

or five decades. P. M. Brown and R. Wu, "Climate and Disturbance Forcing of Episodic Tree Recruitment in a Southwestern Ponderosa Pine Landscape," *Ecology* 86, no. 11 (2005): 3030–38. Also Lacey E. Hankin et al. "Impacts of Growing-Season Climate on Tree Growth and Post-Fire Regeneration in Ponderosa Pine and Douglas-Fir Forests." *Ecosphere* (2019): 10(4):e02679. 10.1002/ecs2.267.

48 *our reforestation shifted and really slowed and went to following major fires.* Walsh coauthored a paper with lots more detail about her plans for the King burn: Malcolm P. North et al., "Tamm Review: Reforestation for Resilience in Dry Western US Forests," *Forest Ecology and Management* 432 (2019): 209–24.

49 *chaparral . . . a serious, long-lasting obstacle to conifers, especially within large patches of high-severity fire.* C. A. Lauvaux, C. N. Skinner, and A. H. Taylor, "High Severity Fire and Mixed Conifer Forest-Chaparral Dynamics in the Southern Cascade Range, USA," *Forest Ecology and Management* 363 (2016): 74–85.

50 *The 2002 Hayman fire was Colorado's largest and most severe.* Paula J. Fornwalt et al., "Did the 2002 Hayman Fire, Colorado, USA, Burn with Uncharacteristic Severity?," *Fire Ecology* 12, no. 3 (2016): 117–32, doi.org/10.4996/fireecology.1203117. Ten years after the Hayman fire, it appears that enormous high-severity burn areas may remain deforested for a very long time; but if you want to look at a bright side, 89 percent of species are native herbs and shrubs, and total plant cover is higher than before the fire. S. R. Abella and P. J. Fornwalt, "Ten Years of Vegetation Assembly After a North American Mega Fire," *Global Change Biology* 21, no. 2 (2015): 789–802.

50 *Apparently conifer seedlings in the Klamaths can hold their own.* Alan J. Tepley et al., "Vulnerability to Forest Loss Through Altered Postfire Recovery Dynamics in a Warming Climate in the Klamath Mountains," *Global Change Biology* 23, no. 10 (2017): 4117–32.

51 *a climate expected to convert one-third of the Klamath region's conifer forest to broadleaf.* J. M. Serra-Diaz et al., "Averaged 30-Year Climate Change Projections Mask Opportunities for Species Establishment," *Ecography* 39 (2016): 844–45.

51 *Fire in the West is on the rise. The increase since 1980 has been hard to miss.* John T. Abatzoglou and A. Park Williams, "Impact of Anthropogenic Climate Change on Wildfire Across Western US Forests," *Proceedings of the National Academy of Sciences of the United States of America* 113, no. 42 (October 2016): 11770–75. Also Philip E. Dennison et al., "Large Wildfire Trends in the Western United States, 1984–2011," *Geophysical Research Letters* 41, no. 8 (2014): 2928–33.

52 *high-severity fire) has increased in California.* C. Mallek et al., "Modern Departures in Fire Severity and Area Vary by Forest Type, Sierra Nevada and Southern Cascades, California, USA," *Ecosphere* 4, no. 12 (2013): 1–28.

52 *it is rebutted in some areas by more reliable data.* Brandon M. Collins et al., "A Quantitative Comparison of Forest Fires in Central and Northern California Under Early (1911–1924) and Contemporary (2002–2015) Fire Suppression," *International Journal of Wildland Fire* 28, no. 2 (2019): 138–48, doi .org/10.1071/WF18137.

52 *What's ominous ecologically is the increasing area in large deforested patches.* Jens T. Stevens et al., "Changing Spatial Patterns of Stand-Replacing Fire in California Conifer Forests," *Forest Ecology and Management* 406 (2017): 28–36. Also Brian J. Harvey, D. C. Donato, and M. G. Turner, "Drivers and Trends in Landscape Patterns of Stand-Replacing Fire in Forests of the US Northern Rocky Mountains (1984–2010)," *Landscape Ecology* 31, no. 10 (2016): 2367–83. Also C. Alina Cansler and Donald McKenzie, "Climate, Fire Size, and Biophysical Setting Control Fire Severity and Spatial Pattern in the Northern Cascade Range, USA," *Ecological Applications* 24, no. 5 (July 2014): 1037–56. Also Keala Hagmann and Andrew Merschel, "Stand-Replacing Fire in Historically Frequent-Fire Forests in South-Central Oregon," abstract of presentation at AFE Fire Continuum Conference, Missoula, MT, May 24, 2018. Also Paul Hessburg, N. Povak, and R. B. Salter, "Early Successional Conditions in the Eastern Washington Cascade Mountains: Contrasting the Pre-management and Modern Eras," abstract of presentation at AFE Fire Continuum Conference, Missoula, MT, May 24, 2018.

52 *With warming, all three trends in spruce-fir forest.* Winslow D. Hansen et al. "Can Wildland Fire Management Alter 21st-Century Subalpine Fire and Forests in Grand Teton National Park, Wyoming, USA?" *Ecological Applications* (2019) doi: 10.1002/eap.2030.

52 *the proportion of those that do not appear to be regenerating as forest.* C. S. Stevens-Rumann et al., "Evidence for Declining Forest Resilience to Wildfires under Climate Change," *Ecology Letters* 21, no. 2 (2018): 243–52, doi .org/10.1111/ele.12889.

52 *That graph is based on crude data from the National Interagency Fire Center, but the Center disavows it.* "Total Wildland Fires and Acres (1926–2017)," National Interagency Coordination Center, National Interagency Fire Center, www.nifc.gov/fireInfo/fireInfo_stats_totalFires.html. Also Zeke Hausfather, "Factcheck: How Global Warming Has Increased US Wild-

fires," CarbonBrief, last modified August 9, 2018, www.carbonbrief.org /factcheck-how-global-warming-has-increased-us-wildfires.

52 *In contrast, there's an excellent three-thousand-year fire history of the West.* Jennifer R. Marlon et al., "Long-Term Perspective on Wildfires in the Western USA," *Proceedings of the National Academy of Sciences* 109, no. 9 (2012): E535–E543.

53 *it's a deficit specifically of low-severity fire.* C. Mallek et al., "Modern Departures in Fire Severity and Area Vary by Forest Type, Sierra Nevada and Southern Cascades, California, USA," *Ecosphere* 4, no. 12 (2013): 1–28. Ryan D. Haugo et al., "The Missing Fire: Quantifying Human Exclusion of Wildfire in Pacific Northwest Forests, USA," *Ecosphere* 10, no. 4 (2019): e02702.

53 *fires in increasing number and size are not optional. They are going to happen.* Anthony LeRoy Westerling, "Increasing Western US Forest Wildfire Activity: Sensitivity to Changes in the Timing of Spring," *Philosophical Transactions of the Royal Society of London. Society B: Biological Sciences* 371, no. 1696 (June 2016): 20150178.

53 *In projected near-term warmer climates they are much less vulnerable to either fire or drought.* Polly C. Buotte et al., "Near-Future Forest Vulnerability to Drought and Fire Varies Across the Western United States," *Global Change Biology* 25, no. 1 (2018): 1–14, doi.org/10.1111/gcb.14490.

53 *at least one study describes microclimate mechanisms that may leverage sweeping ecological change.* Kimberley T. Davis, et al., "Microclimatic Buffering in Forests of the Future: The Role of Local Water Balance," *Ecography* 42, no. 1 (2019): 1–11.

53 *Their huge severe fires of the early 1900s were incubated in clear-cuts.* Lori D. Daniels and R. W. Gray, "Disturbance Regimes in Coastal British Columbia," *BC Journal of Ecosystems and Management* 7 , no. 2 (2006): 44–56.

53 *Fire suppression has not done major harm.* J. S. Halofsky et al., "The Nature of the Beast: Examining Climate Adaptation Options in Forests with Stand-Replacing Fire Regimes," *Ecosphere* 9, no. 3 (2018): e02140, doi.org/10.1002 /ecs2.2140.

53 *These are among the very best forests in the world at sequestering carbon.* B. E. Law et al., "Land Use Strategies to Mitigate Climate Change in Carbon Dense Temperate Forests," *Proceedings of the National Academy of Sciences* 115, no. 14 (2018): 3663–68, doi.org/10.1073/pnas.1720064115.

54 *maximizing their carbon storage by logging them at strictly sustainable rates.* Neil

G. Williams and M. D. Powers, "Carbon Storage Implications of Active Management in Mature *Pseudotsuga menziesii* Forests of Western Oregon," *Forest Ecology and Management* 432 (2019): 761–75. Also K. E. Skog and G. A. Nicholson, "Carbon Sequestration in Wood and Paper Products," in *The Impact of Climate Change on America's Forests*, eds. L. A. Joyce and R. Birdsey (Fort Collins, CO: USDA Forest Service, 2000), 79–88.

55 *We don't like the massive pulses of erosion that typically follow large high-severity fires.* B. P. Murphy, L. L. Yocom, and P. Belmont, "Beyond the 1984 Perspective: Narrow Focus on Modern Wildfire Trends Underestimates Future Risks to Water Security," *Earth's Future* 6, no. 11 (2018): 1492–97, doi .org/10.1029/2018F.F001006.

56 *Charcoal from fires sticks around longer.* Matthew W. Jones et al. "Global Fire Emissions Buffered by the Production of Pyrogenic Carbon." *Nature Geoscience* 12.9 (2019): 742–747.

56 *Satellite imagery's big picture confirms that the world is getting leafier.* Zaichun Zhu et al., "Greening of the Earth and Its Drivers," *Nature Climate Change* 6, no. 8 (2016): 791–95.

56 *Western pine forests haven't been doing as well as some forests globally.* Justin S. Mankin et al., "The Curious Case of Projected Twenty-First-Century Drying but Greening in the American West," *Journal of Climate* 30, no. 21 (2017): 8689–710.

57 *British Columbia's forests were calculated to be a net carbon source.* W. A. Kurz et al., "Mountain Pine Beetle and Forest Carbon Feedback to Climate Change," *Nature* 452, no. 7190 (2008): 987–90. Kurz is also a coauthor of the following hopeful prediction of a renewed carbon sink: V. K. Arora et al., "Potential Near-Future Carbon Uptake Overcomes Losses from a Large Insect Outbreak in British Columbia, Canada," *Geophysical Research Letters* 43, no. 6 (2016): 2590–98, doi.org/10.1002/2015GL067532.

57 *California has also likely become a carbon source since 2000.* Patrick Gonzalez et al., "Aboveground Live Carbon Stock Changes of California Wildland Ecosystems, 2001–2010," *Forest Ecology and Management* 348 (2015): 68–77.

Chapter 3: Outbreak

59 *Most of the lodgepole pine forests of Colorado.* Annie Proulx, *Bird Cloud: A Memoir of Place* (New York: Scribner, 2011), 35.

60 *killing more than half of the pine timber volume in the entire province.* A. Dhar, L. Parrott, and C. Hawkins, "Aftermath of Mountain Pine Beetle Outbreak

in British Columbia: Stand Dynamics, Management Response and Ecosystem Resilience," *Forests* 7, no. 8 (2016): 171.

61 *Whether that will actually happen is the subject of much debate among scientists.* N. C. Coops, M. A. Wulder, and R. H. Waring, "Modeling Lodgepole and Jack Pine Vulnerability to Mountain Pine Beetle Expansion into the Western Canadian Boreal Forest," *Forest Ecology and Management* 274 (2012): 161–71. Also D. W. Goodsman et al., "The Effect of Warmer Winters on the Demography of an Outbreak Insect Is Hidden by Intraspecific Competition," *Global Change Biology* 24, no. 8 (August 2018): 3620–28, doi.org/10.1111/gcb.14284.

61 *As soon as it can decide on a good tree to make a home in.* K. F. Raffa, J. C. Gregoire, and B. S. Lindgren, "Natural History and Ecology of Bark Beetles," in *Bark Beetles: Biology and Ecology of Native and Invasive Species*, eds. F. E. Vega and R. W. Hofstetter (Cambridge, MA: Academic Press, 2015), 1–40. The most recent of many good summaries of bark beetles.

66 *connecting the unprecedented size and location of the recent beetle epidemics to the warming climate is something we can do.* A. S. Weed, M. P. Ayres, and J. A. Hicke, "Consequences of Climate Change for Biotic Disturbances in North American Forests," *Ecological Monographs* 83, no. 4 (2013): 441–70.

67 *Those efforts often fail. It's next to impossible to stay ahead of the beetles.* N. E. Gillette et al., "The Once and Future Forest: Consequences of Mountain Pine Beetle Treatment Decisions," *Forest Science* 60, no. 3 (2014): 527–38.

67 *Early in the twentieth century, foresters were desperate for a solution.* J. M. Miller and F. P. Keen, *Biology and Control of the Western Pine Beetle: A Summary of the First Fifty Years of Research* (Washington, D.C.: U.S. Department of Agriculture, 1960), 275–93.

68 *methylcyclohexenone, and has proven itself in packet form and also in some aerial spraying trials.* S. J. Seybold et al., "Management of Western North American Bark Beetles with Semiochemicals," *Annual Review of Entomology* 63 (2018): 407–32.

71 *Typically, logging companies choose to clear-cut when they salvage log damaged pines.* Larry Pynn, "Pine Beetles: The Aftermath," *Vancouver Sun*, December 2011. Also A. Dhar, L. Parrott, and C. Hawkins, "Aftermath of Mountain Pine Beetle Outbreak in British Columbia: Stand Dynamics, Management Response and Ecosystem Resilience," *Forests* 7, no. 8 (2016): 171.

72 *Big portions of the Yellowstone ecosystem might convert.* Anthony LeRoy Westerling et al., "Continued Warming Could Transform Greater Yellowstone Fire Regimes by Mid-21st Century," *Proceedings of the National Academy of Sciences* 108, no. 32 (2011): 13165–70. So far, no severe reburns at Yellowstone have

come along before the lodgepoles had time to produce new cones; however, severe reburns sixteen and twenty-eight years after a preceding fire have been studied. Lodgepole pine seedlings did come up in these burns, though at only a small fraction of the seedling density after the 1988 fires. Turner, Monica G., et al., "Short-Interval Severe Fire Erodes the Resilience of Subalpine Lodgepole Pine Forests," *Proceedings of the National Academy of Sciences* 116, no. 23 (2019): 11319–28, doi/10.1073/pnas.1902841116. Winslow D. Hansen et al. "Can Wildland Fire Management Alter 21st-Century Subalpine Fire and Forests in Grand Teton National Park, Wyoming, USA?" *Ecological Applications* (2019) doi: 10.1002/eap.2030.

72 *this effect . . . has been reported in similar jack pines in northern Alberta.* B. D. Pinno, R. C. Errington, and D. K. Thompson, "Young Jack Pine and High Severity Fire Combine to Create Potentially Expansive Areas of Understocked Forest," *Forest Ecology and Management* 310 (2013): 517–22.

73 *scientists easily found plots where literally all of the ponderosa pines were dead.* C. J. Fettig et al., "Tree Mortality Following Drought in the Central and Southern Sierra Nevada, California, US," *Forest Ecology and Management* 432 (2019): 164–78.

73 *Bob Van Pelt described five rivals for the title of Champion Sugar Pine.* Robert Van Pelt, *Forest Giants of the Pacific Coast* (Seattle: University of Washington Press, 2001), 72–79.

74 *the driest four-year stretch in the past thousand years, by some calculations.* D. Griffin and K. J. Anchukaitis, "How Unusual Is the 2012–2014 California Drought?," *Geophysical Research Letters* 41, no. 24 (2014): 9017–23, doi .org/10.1002/2014GL062433.

75 *Foresters predicted this epidemic back in 1995.* W. W. Oliver, "Is Self-Thinning in Ponderosa Pine Ruled by *Dendroctonus* Bark Beetles?," in *Forest Health Through Silviculture*, comp. L. G. Eskew, proceedings of the 1995 National Silviculture Workshop, GTR-RM-267, USDA Forest Service, 213–18.

75 *To figure out exactly how this beetle explosion correlated with the drought.* Derek J. N. Young et al., "Long-Term Climate and Competition Explain Forest Mortality Patterns under Extreme Drought," *Ecology Letters* 20, no. 1 (2017): 78–86. Also see corroborating results using other methods in M. L. Goulden and R. C. Bales, "California Forest Die-Off Linked to Multi-year Deep Soil Drying in 2012–2015 Drought," *Nature Geoscience* (2019) doi.org/10.1038 /s41561-019-0388-5.

75 *Greg Asner came up with similar answers from more direct measurement.* G. P.

Asner et al., "Progressive Forest Canopy Water Loss During the 2012–2015 California Drought," *Proceedings of the National Academy of Sciences* 113, no. 2 (2016): E249–E255.

76 *he calculates they will represent an unprecedented fuel load.* Scott L. Stephens et al., "Drought, Tree Mortality, and Wildfire in Forests Adapted to Frequent Fire," *BioScience* 68, no. 2 (2018): 77–88.

78 *one fire in central Oregon spawned two studies that came to opposite conclusions.* M. C. Agne, T. Woolley, and S. Fitzgerald, "Fire Severity and Cumulative Disturbance Effects in the Post-Mountain Pine Beetle Lodgepole Pine Forests of the Pole Creek Fire," *Forest Ecology and Management* 366 (2016): 73–86. Also T. Ryan McCarley et al., "Multi-Temporal LiDAR and Landsat Quantification of Fire-Induced Changes to Forest Structure," *Remote Sensing of Environment* 191 (2017): 419–32.

78 *Studies analyzing the data for broad effects of beetle-killed trees on either fire extent or fire severity.* Garrett W. Meigs et al., "Do Insect Outbreaks Reduce the Severity of Subsequent Forest Fires?," *Environmental Research Letters* 11, no. 4 (2016): 045008. Also Sarah J. Hart et al., "Area Burned in the Western United States Is Unaffected by Recent Mountain Pine Beetle Outbreaks," *Proceedings of the National Academy of Sciences of the United States of America* 112, no. 14 (April 2015): 4375–80.

79 *the only tinderbox study based on actually looking at and measuring wildfires.* Daniel D. B. Perrakis et al., "Modeling Wildfire Spread in Mountain Pine Beetle-Affected Forest Stands, British Columbia, Canada," *Fire Ecology* 10, no. 2 (2014): 10–35.

79 *intensity and impact of the fire were anywhere from two to eleven times greater.* Lori D. Daniels, interview, June 19, 2018.

79 *beetle kill in B.C. definitely increased the probability of burning.* Daniel Perrakis, personal communication, November 8, 2019, regarding a study (lead author Chris Stockdale) that was complete, but not yet published.

80 *Scott Stephens's warning concerned a phase that's still in the future.* Scott L. Stephens et al., "Drought, Tree Mortality, and Wildfire in Forests Adapted to Frequent Fire," *BioScience* 68, no. 2 (2018): 77–88.

80 *In at least one case in the Sierra Nevada.* C. I. Millar and D. L. Delany. "Interaction between mountain pine beetle-caused tree mortality and fire behavior in subalpine whitebark pine forests, eastern Sierra Nevada, CA; Retrospective observations." *Forest Ecology and Management* 447 (2019) 195–202.

81 *the question of whether we can reduce the beetle risk by reducing forest density.* C. J. Fettig et al., "Cultural Practices for Prevention and Mitigation of Moun-

tain Pine Beetle Infestations," *Forest Science* 60, no. 3 (2014): 450–63, doi
.org/10.5849/forsci.13-032. Also Diana L. Six, Eric Biber, and Elisabeth
Long, "Management for Mountain Pine Beetle Outbreak Suppression: Does
Relevant Science Support Current Policy?," *Forests* 5, no. 1 (2014): 103–33.
Also C. J. Fettig et al., "A Comment on 'Management for Mountain Pine Bee-
tle Outbreak Suppression: Does Relevant Science Support Current Policy?',"
Forests 5, no. 4 (2014): 822–26, doi.org/10.3390/f5040822. Also Nancy E.
Gillette et al., "The Once and Future Forest: Consequences of Mountain Pine
Beetle Treatment Decisions," *Forest Science* 60, no. 3 (2014): 527–38.

82 *Another intriguing possibility is a direct effect of fire on pitch production in pon-
derosa.* Sharon Hood et al., "Low-Severity Fire Increases Tree Defense Against
Bark Beetle Attacks," *Ecology* 96, no. 7 (2015): 1846–55. Also S. M. Hood,
S. Baker, and A. Sala, "Fortifying the Forest: Thinning and Burning Increase
Resistance to a Bark Beetle Outbreak and Promote Forest Resilience," *Ecolog-
ical Applications* 26, no. 7 (2016): 1984–2000.

83 *parts of Yosemite National Park that have enjoyed a restored fire regime.* Scott
Stephens, "Wildfires in California: Friend or Foe?," TEDx talk (2019), www
.youtube.com/watch?v=2r7JI6zVwf0.

84 *a striking pattern: the dead ones (killed by beetles) had been faster-growing.* C. I.
Millar, R. D. Westfall, and D. L. Delany, "Response of High-Elevation Limber Pine (*Pinus flexilis*) to Multi-Year Droughts and 20th-Century Warming,
Sierra Nevada, California, USA," *Canadian Journal of Forest Research* 37, no.
12 (2007): 2508–20.

84 *whitebark pine—and she found the exact same pattern.* Constance I. Millar
et al., "Forest Mortality in High-Elevation Whitebark Pine (*Pinus albicaulis*) Forests of Eastern California, USA: Influence of Environmental Context,
Bark Beetles, Climatic Water Deficit, and Warming," *Canadian Journal of
Forest Research* 42, no. 4 (2012): 749–65.

88 *Killed and living trees fell into two distinct lineages.* Diana L. Six, Clare Ver-
gobbi, and Mitchell Cutter, "Are Survivors Different? Genetic-Based Selection
of Trees by Mountain Pine Beetle During a Climate Change-Driven Outbreak
in a High-Elevation Pine Forest," *Frontiers of Plant Science* 9 (July 2018): 993.

88 *that practice has been all too common, especially in British Columbia.* Larry Pynn,
"Pine Beetles: The Aftermath," A *Vancouver Sun* series, December 2011.

Chapter 4: Cookie Cutters

92 *statistically the most impressive nonvolcanic peak within the ruggedest mountain
range.* David Metzler and Edward Earl, "ORS (Spire Measure) Home Page,"

accessed November 26, 2018, www.peaklist.org/spire/index.html; "Some Ruggedness (DRS) Comparisons for Mountain Ranges," accessed November 26, 2018, www.peaklist.org/spire/rug/rugged-ranges.html.

92 *Hozomeen was "an imperturbable surl for cloudburst mist," it was "the Void."* Jack Kerouac, *Desolation Angels* (1965; reprint, New York: Riverhead, 1995), 3–4.

95 *cutting a cookie from one side doesn't unduly risk the tree's life. She tests that hypothesis every five years.* E. K. Heyerdahl and S. J. McKay, "Condition of Live Fire-Scarred Ponderosa Pine Twenty-One Years After Removing Partial Cross-Sections," *Tree-Ring Research* 73, no. 2 (2017): 149–53.

96 *she could identify subtle cohorts intermixed with older survivor trees.* For example, Emily K. Heyerdahl et al., *Multicentury Fire and Forest Histories at 19 Sites in Utah and Eastern Nevada*, Gen. Tech. Rep. RMRS-GTR-261WWW (Fort Collins, CO: USDA Forest Service, 2011). Also E. K. Heyerdahl, R. Loehman, and D. A. Falk, "A Multi-century History of Fire Regimes along a Transect of Mixed-Conifer Forests in Central Oregon, USA," *Canadian Journal of Forest Research* 49, no. 1 (2019): 76–86.

98 *In 1890 there just wasn't much mixed-conifer forest.* Andrew G. Merschel, Thomas A. Spies, and Emily K. Heyerdahl, "Mixed-Conifer Forests of Central Oregon: Effects of Logging and Fire Exclusion Vary with Environment," *Ecological Applications* 24, no. 7 (2014): 1670–88. Also J. D. Johnston, J. D. Bailey, and C. J. Dunn, "Influence of Fire Disturbance and Biophysical Heterogeneity on Pre-settlement Ponderosa Pine and Mixed Conifer Forests," *Ecosphere* 7, no. 11 (2016): e01581.

99 *the original dendrochronologist, Andrew Ellicott Douglass.* D. J. McGraw, "Andrew Ellicott Douglass and the Giant Sequoias in the Founding of Dendrochronology," *Tree-Ring Research* 59, no. 1 (2003): 21–27. Also E. W. Haury, "Recollections of a Dramatic Moment in Southwestern Archaeology," *Tree-Ring Bulletin* 24, nos. 3–4 (1962): 11–14.

101 *lidar aerial images that reveal thousands of collapsed stone habitations.* M. J. Liebmann and T. W. Swetnam et al., "Native American Depopulation, Reforestation, and Fire Regimes in the Southwest United States, 1492–1900 CE," *Proceedings of the National Academy of Sciences* 113, no. 6 (2016): E696–E704.

103 *They found four phases in the fire history here.* Thomas W. Swetnam et al., "Multiscale Perspectives of Fire, Climate and Humans in Western North America and the Jemez Mountains, USA," *Philosophical Transactions of the Royal Society of London, Society B: Biological Sciences* 371, no. 1696 (June 2016): 20150168.

Chapter 5: The Bleeding Edge

110 *The Civil War battle of Glorieta Pass.* "Battle of Glorieta Pass," Pecos National Historical Park, National Park Service, last modified August 8, 2018, www .nps.gov/peco/learn/historyculture/battle-of-glorieta-pass.htm.

113 *In consequence, piñon-juniper savanna was a frequent low-severity fire regime.* E. Q. Margolis, "Fire Regime Shift Linked to Increased Forest Density in a Piñon–Juniper Savanna Landscape," *International Journal of Wildland Fire* 23, no. 2 (2014): 234–45.

113 *savanna is just one of three piñon-juniper community types.* W. H. Romme et al., "Historical and Modern Disturbance Regimes, Stand Structures, and Landscape Dynamics in Piñon–Juniper Vegetation of the Western United States," *Rangeland Ecology and Management* 62, no. 3 (2009): 203–22.

116 *Allen speaks of "hotter droughts" to make the point that this is a clear effect of climate change.* C. D. Allen, D. D. Breshears, and N. G. McDowell, "On Underestimation of Global Vulnerability to Tree Mortality and Forest Die-Off from Hotter Drought in the Anthropocene," *Ecosphere* 6, no. 8 (2015): 129, dx.doi.org/10.1890/ES15-00203.1.

117 *developed computer models and published predictions, and they are very grim.* Nate G. McDowell et al., "Multi-scale Predictions of Massive Conifer Mortality Due to Chronic Temperature Rise," *Nature Climate Change* 6, no. 3 (2016): 295. McDowell later pinned down that pockets of piñon survival would require access to bedrock water and isolation from other piñons attacked by piñon ips. Nate McDowell et al., "Mechanisms of a Coniferous Refugium Persistence Under Drought and Heat," *Environmental Research Letters* 14, no. 4 (2019): doi.org/10.1088/1748-9326/ab0921. Also Darin J. Law et. al., "Bioclimatic Envelopes for Individual Demographic Events Driven by Extremes: Plant Mortality from Drought and Warming," *International Journal of Plant Sciences* 180, no. 1 (2019): 53–62.

118 *half of the park's woodland went up in wildfires. Then between 2002 and 2005, bark beetles.* M. Lisa Floyd, D. D. Hanna, W. H. Romme, "Historical and Recent Fire Regimes in Piñon–Juniper Woodlands on Mesa Verde, Colorado, USA," *Forest Ecology and Management* 198, nos. 1–3 (2004): 269–89. Also M. Lisa Floyd et al., "Structural and Regenerative Changes in Old-Growth Piñon–Juniper Woodlands Following Drought-Induced Mortality," *Forest Ecology and Management* 341 (2015): 18–29.

119 *awaiting the arrival of a new tool that looks like the most promising to date.* Ann C. Kennedy, "Selective Soil Bacteria to Manage Downy Brome, Jointed Goatgrass, and Medusahead and Do No Harm to Other Biota," *Biological Con-*

trol 123 (2018): 18–27. Also Christopher Solomon, "Researcher Finds Way to Fight Cheatgrass, a Western Scourge," *New York Times*, October 8, 2015, nyti .ms/1RrgbQi.

121 *cedar bark beetles are a very significant mortality factor.* Nathan L. Stephenson et al., "Which Trees Die During Drought? The Key Role of Insect Host-tree Selection," *Journal of Ecology* (2019): doi.org/10.1111/1365-2745.13176.

Chapter 6: Thin and Burn

126 *When it's "forest restoration," it aims to leave the stand with a mix of tree sizes.* Jerry F. Franklin et al., *Restoration of Dry Forests in Eastern Oregon: A Field Guide* (Portland, OR: The Nature Conservancy, 2013), 202. Also Paul F. Hessburg et al., "Tamm Review: Management of Mixed-Severity Fire Regime Forests in Oregon, Washington, and Northern California," *Forest Ecology and Management* 366 (2016): 221–50.

127 *These woodlands persevered and thrived for millennia because their fires.* M. Lisa Floyd, D. D. Hanna, W. H. Romme, "Historical and Recent Fire Regimes in Piñon–Juniper Woodlands on Mesa Verde, Colorado, USA," *Forest Ecology and Management* 198, nos. 1–3 (2004): 269–89.

128 *in a position paper from Sierra Pacific Industries.* Cedric Twight, "RE: California Forest Management in Response to Tree Mortality Crisis," letter to Pedro Nava, Chairman, Milton Marks Commission on California State Government Organization and Economy, April 6, 2017, lhc.ca.gov/sites/lhc.ca.gov /files/Reports/242/WrittenTestimony/TwightApr2017.pdf.

128 *Malcolm North, a Forest Service research ecologist, added up.* Malcolm North et al., "Constraints on Mechanized Treatment Significantly Limit Mechanical Fuels Reduction Extent in the Sierra Nevada," *Journal of Forestry* 113, no. 1 (2014): 40–48.

128 *the portion of total area you need to treat.* Mark A. Finney et al., "Simulation of Long-Term Landscape-Level Fuel Treatment Effects on Large Wildfires," *International Journal of Wildland Fire* 16, no. 6 (2007): 712–27.

129 *increasing interest in intentionally shooting for mixed-severity fire.* Van R. Kane et al. "First-Entry Wildfires Can Create Opening and Tree Clump Patterns Characteristic of Resilient Forests," *Forest Ecology and Management* 454 (2019): 117659.

133 *wildfire smoke is likely to be many times worse for us than prescribed fire smoke.* Xiaoxi Liu et al., "Airborne Measurements of Western U.S. Wildfire Emissions: Comparison with Prescribed Burning and Air Quality Implications,"

Journal of Geophysical Research: Atmospheres 122, no. 11 (2017): 6108–29, doi
.org/10.1002/2016JD026315.

134 *Elsewhere, studies find that as trees get more drought-stressed.* Philip J. van Mant-
gem et al., "Pre-fire Drought and Competition Mediate Post-fire Conifer
Mortality in Western U.S. National Parks," *Ecological Applications* 28, no. 7
(2018): 1730–39.

135 *B.C. Forestry oversees three-fourths as much acreage as the U.S. Forest Service.*
Ben Parfitt, *Axed: A Decade of Cuts to BC's Forest Service*, Canadian Centre for
Policy Alternatives, Sierra Club of BC, 2011.

135 *preponderance of evidence does not support that view.* E. L. Kalies and L. L. Yocom Kent, "Are Fuel Treatments Effective at Achieving Ecological and Social
Objectives? A Systematic Review," *Forest Ecology and Management* 375 (2016):
84–95.

135 *huge differences among . . . fuel treatment prescriptions.* Morris C. Johnson
and Maureen C. Kennedy, "Altered Vegetation Structure From Mechanical
Thinning Treatments Changed Wildfire Behaviour in the Wildland-Urban
Interface on the 2011 Wallow Fire, Arizona, USA," *International Journal of
Wildland Fire* 28, no. 3 (2019): 216–29, doi.org/10.1071/WF18062.

135 *fuel treatments make a bigger difference in the part of the fire that burns in extreme.*
John P. Roccaforte et al., "Delayed Tree Mortality, Bark Beetle Activity, and
Regeneration Dynamics Five Years Following the Wallow Fire, Arizona, USA,"
Forest Ecology and Management 428 (2018): 20–26. Also Jamie M. Lydersen,
"Evidence of Fuels Management and Fire Weather Influencing Fire Severity in
an Extreme Fire Event," *Ecological Applications* 27, no. 7 (2017): 2013–30.

136 *In the big blow-up part of the Las Conchas fire, fire was significantly less severe.* R. B. Walker et al., "Fire Regimes Approaching Historic Norms Reduce
Wildfire-Facilitated Conversion from Forest to Non-forest," *Ecosphere* 9, no. 4
(2018): e02182, doi.org/10.1002/ecs2.2182.

136 *During a single nine-hour stretch that reached ninety-four degrees Farenheit . . . it
grew by 192 square miles.* Peter H. Morrison, "Carlton Complex Wildfires—A
Rapid Initial Assessment of the Impact of Washington State's Largest Wildfire" (Winthrop, WA: Pacific Biodiversity Institute, 2014).

136 *Past fuel treatments tempered the fires modestly but significantly that day.* Susan
Prichard, personal communication.

138 *An invasive weedy shrub, Scotch broom.* Zeke Lunder, Facebook, November 10,
2018, 9:14 p.m.

139 *Jon Keeley, warns that if chaparral burns at, say, ten-year intervals.* Jon E. Keeley

and Teresa J. Brennan, "Fire-Driven Alien Invasion in a Fire-Adapted Ecosystem," *Oecologia* 169, no. 4 (2012): 1043–52.

140 *where are the non-fire-prone places in California?* Alistair Smith and Crystal Kolden, "How to Survive Wildfires: Let's Copy Tactics from Nature," *Guardian*, November 14, 2018, www.theguardian.com/environment/2018/nov/14/wildfire-survival-california-nature-trees.

144 *Beverly Law and her lab at Oregon State.* B. E. Law et al., "Land Use Strategies to Mitigate Climate Change in Carbon Dense Temperate Forests," *Proceedings of the National Academy of Sciences* 115 no. 14 (2018): 3663–68, doi .org/10.1073/pnas.1720064115.

145 *Modern pellet stoves produce more heat.* David Bowman et al., "Can Air Quality Management Drive Sustainable Fuels Management at the Temperate Wildland–Urban Interface?," *Fire* 1, no. 2 (2018): 27.

145 *Keane's study saw very little decay.* Robert E. Keane et al., "Surface Fuel Characteristics, Temporal Dynamics, and Fire Behavior of Masticated Mixed-Conifer Fuelbeds of the US Southeast and Rocky Mountains," Final Report, JFSP Project ID: 13-1-05-8, 2017, www.firescience.gov.

146 *Simple, low-budget new ways of turning slash piles into biochar.* Sally James, "Turning Slash Piles into Soil Benefit," *UW News*, October 6, 2011, www .washington.edu/news/2011/10/06/turning-slash-piles-into-soil-benefit.

147 *Mitch Friedman.* M. Friedman, "The Forest Service Is Dead: Long Live the Forest Service!" *Grist Magazine*, February 28, 2006.

147 *Pyrolysis units make biochar while they also produce oil.* Heather G. Wise, Anthony B. Dichiara, and Fernando Resende. "Ex-Situ Catalytic Fast Pyrolysis of Beetle-Killed Lodgepole Pine in a Novel Ablative Reactor." *Fuel* 241 (2019): 933–940.

148 *"For decades, forest activists . . ."* Rick Brown, "Getting From 'No' to 'Yes': A Conservationist's Perspective," in *Old Growth in a New World: A Pacific Northwest Icon Reexamined*, eds. Thomas A. Spies and Sally L. Duncan (Washington, D.C.: Island Press, 2012), 154.

149 *Their reconstructions claim that high-severity fires.* For example, M. A. Williams and W. L. Baker, "Spatially Extensive Reconstructions Show Variable-Severity Fire and Heterogeneous Structure in Historical Western United States Dry Forests," *Global Ecology and Biogeography* 21, no. 10 (2012): 1042–52.

149 *could not have remained standing if high-severity fire proportions and patch sizes had always been what they are today.* Jay D. Miller and Hugh D. Safford, "Corroborating Evidence of a Pre-Euro-American Low- to Moderate-Severity Fire

Regime in Yellow Pine-Mixed Conifer Forests of the Sierra Nevada, California, USA," *Fire Ecology* 13, no. 1 (April 1, 2017): 58–90.

149 *detailed rebuttals of the assumptions.* Peter Z. Fulé et al., "Unsupported Inferences of High-Severity Fire in Historical Dry Forests of the Western United States: Response to Williams and Baker," *Global Ecology and Biogeography* 23, no. 7 (2014): 825–30. Also R. Keala Hagmann et al., "Improving the Use of Early Timber Inventories in Reconstructing Historical Dry Forests and Fire in the Western United States: Comment," *Ecosphere* 9, no. 7 (2018): e02232. Also C. R. Levine et al., "Evaluating a New Method for Reconstructing Forest Conditions from General Land Office Survey Records," *Ecological Applications* 27, no. 5 (2017): 1498–1513. Increasingly sophisticated tree ring research also continues to reconfirm low-severity regimes empirically: E. Q. Margolis and S. B. Malevich, "Historical Dominance of Low-Severity Fire in Dry and Wet Mixed-Conifer Forest Habitats of the Endangered Terrestrial Jemez Mountains Salamander (*Plethodon neomexicanus*)," *Forest Ecology and Management* 375 (2016): 12–26. Also J. D. Johnston, J. D. Bailey, and C. J. Dunn, "Influence of Fire Disturbance and Biophysical Heterogeneity on Pre-settlement Ponderosa Pine and Mixed Conifer Forests," *Ecosphere* 7, no. 11 (2016): e01581. Also M. A. Battaglia et al., "Changes in Forest Structure Since 1860 in Ponderosa Pine Dominated Forests in the Colorado and Wyoming Front Range, USA," *Forest Ecology and Management* 422 (2018): 147–60. Andrew G. Merschel et al., "Influence of Landscape Structure, Topography, and Forest Type on Spatial Variation in Historical Fire Regimes, Central Oregon, USA," *Landscape Ecology* 33, no. 7 (2018): 1195–1209.

151 sidebar, *without drought stress, symptoms of pathogens.* Nancy Grulke et al., *Quantitative and Qualitative Approaches to Assess Effectiveness of Fuels Reduction Treatments in Improving Tree and Stand Health in Dry Pine Forests*, manuscript in draft, 2019.

152 *Another argument against fuel treatments that we sometimes hear is about snags.* Richard L. Hutto et al., "Toward a More Ecologically Informed View of Severe Forest Fires," *Ecosphere* 7, no. 2 (2016): e01255.

152 *already three times more snags in the Sierra Nevada than there had been in 1850.* Hugh D. Safford and Jens T. Stevens, "Natural Range of Variation for Yellow Pine and Mixed-Conifer Forests in the Sierra Nevada, Southern Cascades, and Modoc and Inyo National Forests, California, USA," Gen. Tech. Rep. PSW-GTR-256, USDA Forest Service, 2017.

152 *Pro-treatment ecologists also favor leaving snags standing. California is somewhat*

exceptional, though. Malcolm P. North et al., "Tamm Review: Reforestation for Resilience in Dry Western US Forests," *Forest Ecology and Management* 432 (2019): 209–24.

154 *In Washington's 2006 Tripod fire, thirty-year-old fuels treatments reduced fire.* S. J. Prichard and M. C. Kennedy, "Fuel Treatments and Landform Modify Landscape Patterns of Burn Severity in an Extreme Fire Event," *Ecological Applications* 24, no. 3 (2014): 571–90.

154 *once we've treated we have to allow wildfires to come in.* K. Barnett et al., "Beyond Fuel Treatment Effectiveness: Characterizing Interactions Between Fire and Treatments in the US," *Forests* 7, no. 10 (2016): 237.

154 *analysis finds that current treatments and fires meet 40 to 45 percent of that goal.* Nicole M. Vaillant and Elizabeth D. Reinhardt, "An Evaluation of the Forest Service Hazardous Fuels Treatment Program—Are We Treating Enough to Promote Resiliency or Reduce Hazard?," *Journal of Forestry* 115, no. 4 (2017): 300–8.

155 *Whereas buffers may improve our fire suppression effectiveness.* E. D. Reinhardt et al., "Objectives and Considerations for Wildland Fuel Treatment in Forested Ecosystems of the Interior Western United States," *Forest Ecology and Management* 256, no. 12 (2008): 1997–2006.

155 *Thinning treatments can be "anchors" that facilitate reintroducing fire.* Carmen L. Tubbesing et al., "Strategically Placed Landscape Fuel Treatments Decrease Fire Severity and Promote Recovery in the Northern Sierra Nevada," *Forest Ecology and Management* 436 (2019): 45–55. Malcolm North et al., "Constraints on Mechanized Treatment Significantly Limit Mechanical Fuels Reduction Extent in the Sierra Nevada," *Journal of Forestry* 113, no. 1 (2014): 40–48. Also M. P. North et al., "Reform Forest Fire Management," *Science* 349, no. 6254 (2015): 1280–81.

156 *"Wildfire Risk as Socioecological Pathology."* A. P. Fischer et al., "Wildfire Risk as a Socioecological Pathology," *Frontiers in Ecology and the Environment* 14, no. 5 (2016): 276–84.

156 *"You have houses strung all over the country and in some of the most terrible places."* Anonymous interviewee in J. Canton-Thompson et al., "External Human Factors in Incident Management Team Decisionmaking and Their Effect on Large Fire Suppression Expenditures," *Journal of Forestry* 106, no. 8 (2008): 416–24.

156 *I've seen dueling statistics on what percentage of firefighting costs go toward protecting the WUI.* P. H. Gude et al., "Evidence for the Effect of Homes on Wildfire Suppression Costs," *International Journal of Wildland Fire* 22, no. 4 (2013): 537–48. On the other hand, Sarah McCaffrey, "Fire Nar-

ratives: Are Any Accurate?," presentation at AFE Fire Congress, filmed 2017, Orlando, FL, video, 19:49, mediasite.video.ufl.edu/Mediasite/Play/4ee7b1c5adeb4d3397c55670f44cb72c1d.

157 *shut down . . . simply because too many people complained about smoke.* Julie Cart, "'Environmental Catastrophe': Can California Play Catch-up on Forest Management to Prevent Future Disastrous Wildfires?," *Chico News and Review*, April 12, 2018, www.newsreview.com/chico/environmental-catastrophe/content?oid=26106279.

158 *support) among WUI residents for fuel treatments, prescribed burns, and even managed wildfire.* S. M. McCaffrey and C. S. Olsen, "Research Perspectives on the Public and Fire Management: A Synthesis of Current Social Science on Eight Essential Questions," Gen. Tech. Rep. NRS-104, Northern Research Station, USDA Forest Service, 2012.

158 *The state's biggest historical fire, the Canyon Creek, ran twenty-one miles.* M. E. Alexander, *The 1988 Fires of Yellowstone and Beyond as a Wildland Fire Behavior Case Study*, Wildland Fire Lessons Learned Center, USDA Forest Service, 2009.

159 *Two fires, it turns out, did a great job of restoration.* Andrew J. Larson et al., "Latent Resilience in Ponderosa Pine Forest: Effects of Resumed Frequent Fire," *Ecological Applications* 23, no. 6 (2013): 1243–49. Also C. Stevens-Rumann and P. Morgan, "Repeated Wildfires Alter Forest Recovery of Mixed-Conifer Ecosystems," *Ecological Applications* 26, no. 6 (2016): 1842–53.

Chapter 7: North, and Up

163 *for most trees in the West Coast states, the total range where seedlings are growing is slightly colder.* Vicente J. Monleon and Heather E. Lintz, "Evidence of Tree Species' Range Shifts in a Complex Landscape," *PLoS One* 10, no. 1 (2015): e0118069.

164 *At around twelve thousand years ago it was shockingly rapid—like sixteen degrees Fahrenheit of warming in fifty years.* John W. Williams and Kevin D. Burke, "Past Abrupt Changes in Climate and Terrestrial Ecosystems," in *Biodiversity and Climate Change: Transforming the Biosphere*, eds. T. E. Lovejoy and L. Hannah (New Haven: Yale University Press, 2019): 128–41. J. P. Steffensen et al., "High-Resolution Greenland Ice Core Data Show Abrupt Climate Change Happens in Few Years," *Science* 321, no. 5889 (2008): 680–84.

164 *Oregon experienced an abrupt warming roughly simultaneous with Greenland's at 11,700 years ago.* V. Ersek et al., "Holocene Winter Climate Variability in Mid-latitude Western North America," *Nature Communications* 3 (2012): 1219.

165 *The winters were very cold, though, so the climate overall was not a close analogue to our century.* C. Whitlock et al., "Holocene Vegetation, Fire and Climate History of the Sawtooth Range, Central Idaho, USA," *Quaternary Research* 75, no. 1 (2011): 114–24.

165 *Preserved in heavy crusts of ancient dried packrat urine, middens can keep.* S. T. Jackson et al., "A 40,000-Year Woodrat-Midden Record of Vegetational and Biogeographical Dynamics in North-eastern Utah, USA," *Journal of Biogeography* 32, no. 6 (2005): 1085–106.

166 *slow and broken northward progress in the Rockies.* Jodi R. Norris, Julio L. Betancourt, and Stephen T. Jackson, "Late Holocene Expansion of Ponderosa Pine (*Pinus ponderosa*) in the Central Rocky Mountains, USA," *Journal of Biogeography* 43, no. 4 (2016): 778–90.

166 *they undoubtedly were assisted by pinyon jays or Clark's nutcrackers.* M. R. Lesser and S. T. Jackson, "Contributions of Long-Distance Dispersal to Population Growth in Colonising *Pinus ponderosa* Populations," *Ecology Letters* 16, no. 3 (2013): 380–89.

166 *Two-needle piñon pine also spread to most of its current range by six thousand years ago.* S. T. Gray et al., "Role of Multidecadal Climate Variability in a Range Extension of Pinyon Pine," *Ecology* 87, no. 5 (2006): 1124–30.

167 *Lodgepole migrated rapidly following ice melt to fill up its present B.C. distribution.* Mary Edwards et al., "The Role of Fire in the Mid-Holocene Arrival and Expansion of Lodgepole Pine (*Pinus contorta*) in Yukon, Canada," *The Holocene* 25, no. 1 (2015): 64–78.

167 *Mountain hemlock had to fly 150 miles over inhospitably arid country.* Erin M. Herring et al., "Ecological History of a Long-Lived Conifer in a Disjunct Population," *Journal of Ecology* 106, no. 1 (2018): 319–32. Also S. M. Rosenberg, I. R. Walker, and R. W. Mathewes, "Postglacial Spread of Hemlock (*Tsuga*) and Vegetation History in Mount Revelstoke National Park, British Columbia, Canada," *Canadian Journal of Botany*, 81, no. 2 (2003): 139–51.

167 *around Revelstoke there grow rain forests of western hemlock and western redcedar.* Daniel G. Gavin, "The Coastal-Disjunct Mesic Flora in the Inland Pacific Northwest of USA and Canada: Refugia, Dispersal and Disequilibrium," *Diversity and Distributions* 15, no. 6 (2009): 972–82.

167 *Single-leaf piñon pine took all eleven millennia to crawl about two hundred miles.* C. I. Millar and W. B. Woolfenden, "Ecosystems Past: Prehistory of California Vegetation," in eds. E. Zavaleta and H. Mooney, *Ecosystems of California* (Oakland: University of California Press, 2016), 131–54.

168 *By 3,500 years ago the climate was cooling and treelines moved back down.* G.

Jiménez-Moreno and R. S. Anderson, "Pollen and Macrofossil Evidence of Late Pleistocene and Holocene Treeline Fluctuations from an Alpine Lake in Colorado, USA," *The Holocene* 23, no. 1 (2013): 68–77. Also S. Mensing et al., "A 15,000 Year Record of Vegetation and Climate Change From a Treeline Lake in the Rocky Mountains, Wyoming, USA," *The Holocene* 22, no. 7 (2012): 739–48.

174 *Tree planting in B.C. will follow data-driven guidelines.* Greg O'Neill et al., "A Proposed Climate-Based Seed Transfer System for British Columbia," Tech. Rep. 099, Forests, Lands, and NR Operations, 2017, www.for.gov.bc.ca/hfd /pubs/Docs/Tr/Tr099.htm.

175 *She boldly advocated for assisting gene flow in a 2007 article.* C. I. Millar, N. L. Stephenson, and S. L. Stephens, "Climate Change and Forests of the Future: Managing in the Face of Uncertainty," *Ecological Applications* 17, no. 8 (2007): 2145–51.

175 *lodgepole pine seedlings from various provenances were tried at various elevations.* Eric Conlisk et al., "Seed Origin and Warming Constrain Lodgepole Pine Recruitment, Slowing the Pace of Population Range Shifts," *Global Change Biology* 24, no. 1 (2018): 197–211.

176 *plant the forward-looking provenances as pure small stands.* Sally N. Aitken et al., "Adaptation, Migration or Extirpation: Climate Change Outcomes for Tree Populations," *Evolutionary Applications* 1, no. 1 (2008): 95–111.

177 *It is properly classed as "neo-native" there.* C. I. Millar, N. L. Stephenson, and S. L. Stephens, "Climate Change and Forests of the Future: Managing in the Face of Uncertainty," *Ecological Applications* 17, no. 8 (2007): 2145–51.

178 *it would be more cautious to plant a variety of provenances and species.* C. I. Millar, N. L. Stephenson, and S. L. Stephens, "Climate Change and Forests of the Future: Managing in the Face of Uncertainty," *Ecological Applications* 17, no. 8 (2007): 2145–51. Also A. R. Hof, C. C. Dymond, and D. J. Mladenoff, "Climate Change Mitigation Through Adaptation: The Effectiveness of Forest Diversification By Novel Tree Planting Regimes," *Ecosphere* 8, no. 11 (2017): e01981.

Chapter 8: Ghosts

183 *dead whitebark pines already outnumber living ones.* Sara A. Goeking and Deborah Kay Izlar, "*Pinus albicaulis Engelm* (Whitebark Pine) in Mixed-Species Stands throughout Its US Range: Broad-Scale Indicators of Extent and Recent Decline," *Forests* 9, no. 3 (2018): 131.

185 *often grow in multitrunked form—might those simply be seedling clumps all grown up?* Credit Ronald Lanner for first thinking of that connection. Ronald

M. Lanner, "Avian Seed Dispersal as a Factor in the Ecology and Evolution of Limber and Whitebark Pines," in *Proceedings of Sixth North American Forest Biology Workshop* (Edmonton: University of Alberta, 1980), 15–48.

186 *DNA analysis . . . confirmed the hypothesis about clumped stems.* D. F. Tomback and Y. B. Linhart, "The Evolution of Bird-Dispersed Pines," *Evolutionary Ecology* 4, no. 3 (1990): 185–219.

189 *the bears are finding diet substitutes without having to move very far.* F. T. van Manen et al., "Density Dependence, Whitebark Pine, and Vital Rates of Grizzly Bears," *Journal of Wildlife Management* 80, no. 2 (2016): 300–13. That said, there is some concern that seeking the diet substitutes induces bears to wander into more frequent contact with humans, leading to bear mortality. Johnathan Hettinger, "Yellowstone's Grizzlies Wandering Farther from Home and Dying in Higher Numbers," *Inside Climate News*, May 14, 2019, insideclimatenews.org/news/051419/grizzly-bears-killed-climate-change -yellowstone-rocky-mountains-wildlife-survival-whitebark-pine-beetle.

190 *below one thousand cones per hectare the birds get pretty sparse.* S. T. McKinney, C. E. Fiedler, and D. F. Tomback, "Invasive Pathogen Threatens Bird–Pine Mutualism: Implications for Sustaining a High-Elevation Ecosystem," *Ecological Applications* 19, no. 3 (2009): 597–607.

190 *On the contrary, on sites in the southern Sierra Nevada.* Marc D. Meyer et al., "Mortality, Structure, and Regeneration in Whitebark Pine Stands Impacted by Mountain Pine Beetle in the Southern Sierra Nevada," *Canadian Journal of Forest Research* 46, no. 4 (2016): 572–81.

190 *In more scientific terms, we say that climate envelope models.* M. V. Warwell, G. E. Rehfeldt, and N. L. Crookston, "Modeling Contemporary Climate Profiles of Whitebark Pine (*Pinus albicaulis*) and Predicting Responses to Global Warming," in E. M. Goheen and R. A. Sniezko, *Proceedings of the Conference Whitebark Pine: A Pacific Coast Perspective*, Tech. Rec. R6-NR-FHP-2007-01, USDA Forest Service, 2007, 139–42.

191 *A lot of area opens up for them in northern British Columbia.* Sierra C. McLane and S. N. Aitken, "Whitebark Pine (*Pinus albicaulis*) Assisted Migration Potential: Testing Establishment North of the Species Range," *Ecological Applications* 22, no. 1 (2012): 142–53.

191 *In many places it actually responds to a warmer year by growing faster.* C. I. Millar et al., "Response of Subalpine Conifers in the Sierra Nevada, California, U.S.A., to 20th-Century Warming and Decadal Climate Variability," *Arctic, Antarctic, and Alpine Research* 36, no. 2 (2004): 181–200. Also Kimberly M. Carlson, B. Coulthard, and B. M. Starzomski, "Autumn Snowfall Controls

the Annual Radial Growth of Centenarian Whitebark Pine (*Pinus albicaulis*) in the Southern Coast Mountains, British Columbia, Canada," *Arctic, Antarctic, and Alpine Research* 49, no. 1 (2017): 101–13.

191 *Limber pines are currently demonstrating the distance advantage of bird dispersal.* C. I. Millar et al., "Recruitment Patterns and Growth of High-Elevation Pines in Response to Climatic Variability (1883–2013) in the Western Great Basin, USA," *Canadian Journal of Forest Research* 45, no. 10 (2015): 1299–312.

Chapter 9: Fading White

194 *"highest commercial value of any species, wherever found."* F. I. Rockwell, "The White Pines of Montana and Idaho—Their Distribution, Quality, and Uses," *Forestry Quarterly* 9, no. 2 (1911): 219–31.

195 *White pine blister rust reached New England by 1897 and British Columbia by 1921.* B. W. Geils, K. E. Hummer, and R. S. Hunt, "White Pines, *Ribes*, and Blister Rust: A Review and Synthesis," *Forest Pathology* 40, nos. 3–4 (2010): 147–85.

198 *John Muir called them "the best of sweets."* John Muir, *The Mountains of California* (New York: Century Co., 1894), 157.

198 *His journal tells that the only way he could think of to collect seeds was to shoot.* David Douglas, *Journal Kept by David Douglas, 1823–1827* (reprint, New York: Antiquarian Press, 1959), 230.

202 *barely 1 percent of the whitebarks and none of the foxtail pines were infected.* Jonathan Nesmith et al., "Whitebark and Foxtail Pine in Yosemite, Sequoia, and Kings Canyon National Parks: Initial Assessment of Stand Structure and Condition," *Forests* 10, no. 35 (2019): 35, doi.org/10.3390/f10010035.

203 *One tree killed by bark beetles back in the 1600s fell onto talus and is preserved.* Evan R. Larson, S. L. Van De Gevel, and H. D. Grissino-Mayer, "Variability in Fire Regimes of High-Elevation Whitebark Pine Communities, Western Montana, USA," *Ecoscience* 16, no. 3 (2009): 282–98.

203 *commingled residue of pine beetles and whitebark pines in eight-thousand-year-old sediments.* Andrea Brunelle et al., "Holocene Records of Dendroctonus Bark Beetles in High Elevation Pine Forests of Idaho and Montana, USA," *Forest Ecology and Management* 255, nos. 3–4 (2008): 836–46.

203 *Yellowstone's outbreak of 2007–2010 has slacked off a little.* Erin Shanahan et al., "Whitebark Pine Mortality Related to White Pine Blister Rust, Mountain Pine Beetle Outbreak, and Water Availability," *Ecosphere* 7, no. 12 (2016): e011610.

207 *the partner may abandon them if their populations fall too low.* S. T. McKinney,

C. E. Fiedler, and D. F. Tomback, "Invasive Pathogen Threatens Bird–Pine Mutualism: Implications for Sustaining a High-Elevation Ecosystem," *Ecological Applications* 19, no. 3 (2009): 597–607.

Chapter 10: Resistance

210 *The object is to confirm which individual trees carry genes for resistance, and then to use their seeds.* A relatively broad and recent paper on Sniezko's research is R. A. Sniezko and J. Koch, "Breeding Trees Resistant to Insects and Diseases: Putting Theory into Application," *Biological Invasions* 19, no. 11 (2017): 3377–400.

213 *the earliest field trials to date offer grounds for optimism.* Richard A. Sniezko, Jeremy S. Johnson, and Douglas P. Savin. "Assessing the Durability, Stability, and Usability of Genetic Resistance to a Non-Native Fungal Pathogen in Two Pine Species." *Plants, People, Planet* (2019) 00:1–12. DOI: 10.1002/ppp3.49.

214 *endophytic fungi in white pine needles have been identified and correlated with . . . rust resistance.* Lorinda S. Bullington et al., "The Influence of Genetics, Defensive Chemistry and the Fungal Microbiome on Disease Outcome in Whitebark Pine Trees," *Molecular Plant Pathology* 19, no. 8 (February 1, 2018): 1847–58, doi.org/10.1111/mpp.12663.

Chapter 11: The Enduring

218 *a continuous 8,837-year tree ring record.* Recent work, pending further confirmation, extends the continuous record to 10,359 years, and may be able to go to 11,200. M. W. Salzer, C. L. Pearson, and C. H. Baisan, "Dating the Methuselah Walk Bristlecone Floating Chronologies," *Tree-Ring Research* 75, no. 1 (2019): 61–66.

219 *Giant sequoias, for example, have been sacrificing a lot of their needles.* Nathan L. Stephenson et al., "Patterns and Correlates of Giant Sequoia Foliage Dieback During California's 2012–2016 Hotter Drought," *Forest Ecology and Management* 419 (2018): 268–78.

221 *She brought hungry beetles and confined them in "attack boxes."* E. L. Eidson, K. E. Mock, and B. J. Bentz, "Low Offspring Survival in Mountain Pine Beetle Infesting the Resistant Great Basin Bristlecone Pine Supports the Preference-Performance Hypothesis," *PLoS One* 13, no. 5 (2018): e0196732.

222 *Growth has sped up since 1950 in the groves at highest elevations.* M. W. Salzer et al., "Recent Unprecedented Tree-Ring Growth in Bristlecone Pine at the Highest Elevations and Possible Causes," *Proceedings of the National Academy of Sciences* 106, no. 48 (2009): 20348–53.

223 *they leapfrogged up past the pure bristlecone belt and are setting the new, higher treeline.* C. I. Millar et al., "Recruitment Patterns and Growth of High-Elevation Pines in Response to Climatic Variability (1883–2013) in the Western Great Basin, USA," *Canadian Journal of Forest Research* 45, no. 10 (2015): 1299–312.

Chapter 12: Future Forests

228 *These are realities that scientists agree on (while they sometimes disagree on some of the details).* M. A. Moritz et al., "A Statement of Common Ground Regarding the Role of Wildfire in Forested Landscapes of the Western United States," Fire Research Consensus Working Group Final Report, 2018.

229 *two million acres were burned in 16,626 prescribed fires, 14 of which escaped.* Susan Britting, "2012 Escaped Prescribed Fire Review Summary: Lessons from Escaped Prescribed Fires," written testimony to the Little Hoover Commission, California, 2013.

229 *they prevent more smoke than they produce.* J. W. Long, L. W. Tarnay, and M. P. North, "Aligning Smoke Management with Ecological and Public Health Goals," *Journal of Forestry* 116, no. 1 (2017): 76–86.

230 *Clean Air Act regulates smoke produced by humans, while giving a pass to wildfire smoke.* Kirsten H. Engel, "Perverse Incentives: The Case of Wildfire Smoke Regulation," *Ecology Law Quarterly* 40 (2013): 623.

230 *agencies may need to spend large amounts of money on large-scale fuel treatment.* Matthew Thompson and Nathaniel Anderson, "Modeling Fuel Treatment Impacts on Fire Suppression Cost Savings: A Review," *California Agriculture* 69, no. 3 (2015): 164–70. Also M. P. Thompson et al., "Modeling Fuel Treatment Leverage: Encounter Rates, Risk Reduction, and Suppression Cost Impacts," *Forests* 8, no. 12 (2017): 469.

230 *Whether there would be a net improvement to the forest's carbon budget.* E. Louise Loudermilk et al., "Effectiveness of Fuel Treatments for Mitigating Wildfire Risk and Sequestering Forest Carbon: A Case Study in the Lake Tahoe Basin," *Forest Ecology and Management* 323 (2014): 114–25. Matthew D. Hurteau, et al. "Opinion: Managing for Disturbance Stabilizes Forest Carbon." *Proceedings of the National Academy of Sciences* 116, no. 21 (2019): 10193–10195. Or on the other hand, J. Campbell, M. E. Harmon, and S. R. Mitchell, "Can Fuel-Reduction Treatments Really Increase Forest Carbon Storage in the Western US by Reducing Future Fire Emissions?," *Frontiers in Ecology and Environment* 10, no. 2 (2011): 83–90.

231 *higher levels of future wildfire, and accordingly predict greater carbon-budget benefits.* C. H. Carlson, S. Z. Dobrowski, and H. D. Safford, "Variation in Tree Mortality and Regeneration Affect Forest Carbon Recovery Following Fuel Treatments and Wildfire in the Lake Tahoe Basin, California, USA," *Carbon Balance and Management* 7, no. 1 (2012): 7.

232 *Letting fires burn has been the policy for one drainage in Yosemite.* B. M. Collins and S. L. Stephens, "Stand-Replacing Patches within a 'Mixed Severity' Fire Regime: Quantitative Characterization Using Recent Fires in a Long-Established Natural Fire Area," *Landscape Ecology* 25, no. 6 (2010): 927–39.

232 *Zion National Park has let fires burn since 1988.* Peter M. Brown, C. Gentry, and Q. Yao, "Historical and Current Fire Regimes in Ponderosa Pine Forests at Zion National Park, Utah: Restoration of Pattern and Process After a Century of Fire Exclusion," *Forest Ecology and Management* 445 (2019): 1–12, doi .org/10.1016/j.foreco.2019.04.058.

234 *Mayor Walt Cobb of Williams Lake, B.C., told the* Vancouver Sun. Jennifer Saltman, "Wildfire Mitigation on Municipalities' Minds," *Vancouver Sun*, August 24, 2018.

235 *Effective structural improvements are also possible outside the houses themselves.* Alistair M. Smith et al., "The Science of Firescapes: Achieving Fire-Resilient Communities," *Bioscience* 66, no. 2 (2016): 130–46.

Afterword

239 *hundreds of other Western towns that rate just as vulnerable.* Pamela Ren Larson and Dennis Wagner, "Where Will the West's Next Deadly Wildfire Strike?" *Arizona Republic*, 2019. www.azcentral.com/in-depth/news/local /arizona-wildfires/2019/07/22/wildfire-risks-more-than-500-spots-have -greater-hazard-than-paradise/1434502001/ Accessed October 28, 2019.

239 *a California home's chances of surviving a fire.* Alexandra D. Syphard and Jon E. Keeley, "Factors Associated with Structure Loss in the 2013–2018 California Wildfires," *Fire* 2.3 (2019): 49.

242 *plant a trillion trees.* Jean-François Bastin et al., "The Global Tree Restoration Potential," *Science* 365.6448 (2019): 76–79. Gigatonne is a synonym of petagram, another global-scale unit of carbon emissions; they're each 1,000,000,000,000 kilograms.

243 *Just weeks earlier, in a paper signed by a dozen.* Paul F. Hessburg et al., "Climate, Environment, and Disturbance History Govern Resilience of Western North American Forests," *Frontiers in Ecology and Evolution* 7 (2019): 239. The twenty-nine additional authors include "my" consultants Sean Parks, Su-

san Prichard, Malcolm North, Andrew Larson, Craig Allen, Scott Stephens, Camille Stevens-Rumann, Lori Daniels, Derek Churchill, Keala Hagmann, Alina Cansler, and Hugh Safford.

244 *Team leader Thomas Crowther told a reporter.* Aisling Irwin, "The Everything Mapper," *Nature* 573 (2019): 478–481.

244 *A scientist from Congo complained.* Eike Luedeling et al. ,"Forest Restoration: Overlooked Constraints," *Science* (2019) 10.1126/science.aay7988.

244 *zoomed in on familiar New Mexico areas.* Zooming in can be done at www .crowtherlab.com/maps-2. Accessed October 29, 2019.

245 *native forests do a better job of sequestering carbon.* Simon L. Lewis et al., "Restoring Natural Forests Is the Best Way to Remove Atmospheric Carbon," *Nature* 568.25 (2019): 25–28.

245 *bigger trees do a better job than smaller ones.* Nathan L. Stephenson et al., "Rate of Tree Carbon Accumulation Increases Continuously with Tree Size," *Nature* 507 (2014): 90–93 (2014). Also James A. Lutz et al., "Global Importance of Large-Diameter Trees," *Global Ecology and Biogeography* 27 (2018): 849–864.

245 *letting older forests keep growing may do.* Beverly E. Law et al., "Carbon Storage and Fluxes in Ponderosa Pine Forests at Different Developmental Stages," *Global Change Biology* 7.7 (2001): 755–777.

246 *in scientific studies they identify specific benefits.* Chorong Song, Harumi Ikei, and Yoshifumi Miyazaki, "Physiological Effects of Nature Therapy: A Review of the Research in Japan," *International Journal of Environmental Research and Public Health* 13 (2016): 781. Also M. Kusuhara et al., "A Fragrant Environment Containing α-Pinene Suppresses Tumor Growth in Mice by Modulating the Hypothalamus/Sympathetic Nerve/Leptin Axis and Immune System," *Integrative Cancer Therapies* 18 (2019): 1534735419845139.

Air curtain burners, 146

Allen, Craig, 148, 243, 244; about, 31–33; on biomass burning, 148; on drought, 51, 116, 121–22; on fire aftermath, 44, 146

AMAT (Assisted Migration Adaptation Trial), 171, 174

Angora fire (2007), 22

Anthropogenic warming. See Climate change

Asner, Greg, 75–76

Aspen trees, 44

Assisted gene flow, 175–76, 178

Assisted migration, 172–80, 231; King fire, 24; species rescue AM vs. forestry AM, 177

Asymmetric competition, 105

Bark beetles. See also Mountain pine beetle: as agents of natural selection,
83–85, 87–89; benefits, 83–85; change over time, 59–60; control, 67–69, 81–82; defense against cold snaps, 65–66; description, 61; Douglas-fir beetle, 68; life cycle, 61, 66; notable species, 62–63; outbreaks, 65, 72, 116; pheromones, 64–65, 68–69; piñon ips, 116; predators, 67; trees' defense mechanisms, 64

Bastin, Jean-François, 242; the Bastin paper, 242–45

B.C. Forestry, 135, 172–73

Beetle-killed forests, 77, 78–81

Bentz, Barbara, 67, 221–22

Bienz, Craig, 146

Big Burn (1910), 40

Biochar, 146

Bioenergy, 145, 147

Biscuit fire (2002), 44–45

Bristlecone pine, 201–2, 217–25
Broadleaf evergreens, 50–51
Brown, Rick, 148

Camp fire (2018), 136–39, 238–39
Cansler, Alina, 158
Canyon Creek fire (2017), 158
Caples fire (2019), 241
Carbon cycle, 56–57, 144
Carlton Complex megafire (2014), 136
Carr fire (2018), 36–38, 237–38
Catfaces, 94–95
Cavitation, 122
Cerro Grande fire (2000), 30, 34, 229
CFLRP (Collaborative Forest Landscape
    Restoration Program), 140–41, 145
Chaparral, 47, 49, 139
Charcoal, 146–47
Cheatgrass, 119–21
Chetco Bar fire (2017), 213
Clark's nutcracker, 184–88, 190, 207
Clements, Craig, 35–36
Climate change: in the American West,
    15; Bristlecone pine growth rate,
    222–23; effect on beetle epidemics,
    66; post-glacial warming, 163–65;
    shifting habitats, 42; skeptics, 149–
    50. See also Bastin, Jean-François
Comandra blister rust, 197
Community protection, 153–54,
    156–58, 234–35
Convection plumes, 36
Cougar Creek fire (2018), 50
Coulter pine, 46, 73
Crown fire, 24, 30, 79, 127, 149, 163

Dendrochronology, 93–94, 99–100,
    218–20
Desolation Peak, 91–92
Disease resistance, 210, 213–15
Dome fire (1996), 30

Dorena Genetic Resources Center, 209
Dormancy, 122
Douglass, Andrew Ellicott, 99–100
Drought stress: in British Columbia, 51;
    in California, 74–77; weakening
    effect on trees, 15, 121–22, 134

Eastern white pine, 194–95, 196
Embolism, 122
Endophyte inoculation, 214
Episodic regeneration, 45–46
Erosion, 55, 117. See also Watershed
    restoration

Fire: change over time, 52–53; dynamics,
    measuring, 35; exclusion, 25; intensity
    vs. severity, 23; models, 133–34; scars
    (on trees), 10, 94–95; vortices, 36–38
Firestorm, 31
Fire suppression, 155–56, 160, 230, 232
Fire suppression paradox, 26, 155
Floyd-Hanna, Lisa, 127
Forest recovery, 45–46. See also
    Replanting and reseeding
Forest restoration, 22, 126–27. See
    also Fuel reduction treatments;
    Prescribed burning; Thinning
Forthofer, Jason, 37
4FRI (Four Forest Restoration
    Initiative), 141–43
Foxtail pine, 201–2
Franklin, Jerry Forest, 12–16
Frequent-fire forests, 96, 126, 228
Friedman, Mitch, 147–48
Fuel reduction treatments, 22, 126–27,
    154, 155. See also Prescribed
    burning; Thinning
Fuel-seeking fire, 79

Gambel oak, 31, 43, 45, 50
"Gang of Four," 13

Gasification, 147

Genetic engineering, 214

Grand fir, 98–99

Granite Gulch fire (2019), 241

Grassland fire, 135

Gray-stage fuels, 80

Grazing, 116–17

Great Basin bristlecone pine, 217–25

Greene, Greg, 39, 69, 104–6

Hanceville/Riske Creek fire (2017), 39, 40

Harlan, Tom, 219–20

Hayman fire (2002), 31, 50, 135

Healthy forest, 150–51

Healthy Forests Restoration Act, 150

Heyerdahl, Emily, 93–94, 96–99

High-grade logging, 105, 142

High-severity fire, 23

H. J. Andrews Experimental Forest, 13–15

Hood, Sharon, 82–83

Hygrophobic soils, 44

Incense cedar, 19, 74

Jeffrey pine, 19–20

Jemez Mountains, 101–3

Johnson, Norm, 12–13, 15

Juniper, 110, 114

Keane, Bob, 145–46

Kennedy, Ann, 120

Kerouac, Jack, 92–93

King fire (2014), 23-25, 47–48, 152

Krummholz, 205-207

Ladder fuels, 12

Lake Tahoe, 16–17, 19–20

Larson, Andrew, 158–60

Las Conchas fire (2011), 29–30, 31, 33, 35, 45, 136

Limber pine, 168, 200–201, 223

Lodgepole pine, 69–72, 167

Logging. See Timber industry

Long fire, 22–23, 25

Low-severity fire, 18

Maher, Colin, 204–6

Margolis, Ellis, 140, 142; about, 110–14; fire scar records, 103; future of P-J landscapes, 116–17; prescribed burns, 125–26, 134

Marlon, Jennifer, 52–53

Mass fire, 31

Mastication, 145–46

Mast years, 46, 115, 189

McDowell, Nate, 117

Megafires, 135

Meigs, Garrett, 78, 80–81

Merschel, Andrew, 97–99

Methuselah, 219

Migration of tree species, 163–70

Millar, Connie, 84–85, 89, 175–76

Mixed-severity fire, 96

Mountain hemlock, 167

Mountain pine beetle, 60–62, 65, 203, 221–22. See also Bark beetles

Natural selection, 211

Nelson, Kellen, 132

Niche, 172–73

North, Malcolm, 128, 178

Northwest Forest Plan, 13

Old growth, 13–14

Old-stage fuels, 80

O'Neill, Greg, 170–75

Palynology, 165

Parks, Sean, 26

Parsons, Russ, 131–32

Partial dieback, 218–19

Patchiness: as a goal in replanting, 24; historical practices, 102; importance, 96, 160; lack of, 66–67

Perrakis, Daniel, 79–81, 241

Phenotypic plasticity, 228

Photosynthesis, 121–22

Pine trees, about, 6–7

Piñon-juniper communities: savanna, 111–13, 127; shrublands, 113–14; woodland, 113, 115, 117–18, 127

Piñon pine, 114, 166, 168

Pinyon jay, 114–15

Pioneer fire (2016), 36

Pitch, 64, 82–83

Plateau fire complex (2018), 38–40

Pollen flow, 168–69

Ponderosa pine: description, 9–10; drought resistance, 43; effect on fire, 10–11; forests, importance of, 25; migration, 166

Port Orford cedar, 212

Prescribed burning: compared to thinning, 127–29; cost, 230; effects, 22, 126; efficacy, 135–36, 229, 233; methods, 125; public support, 34

Provenance trial, 171, 174

Proximate agent, 121

Pyrocumulonimbus clouds, 36, 39

Pyrolysis, 147

RaDFIRE (Rapid Deployments to Wildfires Experiment), 35–37

Ramirez, Aaron, 134

Red-stage fuels, 79, 80

Replanting and reseeding, 46–48; among snags, 152–53; of disease-resistant trees, 210; natural processes, 24, 41–42, 187–88; policies, 175

Resilient forest, 151

Restoration harvest, 13, 230

Ribes (genus), 195–96

Rim megafire (2013), 133, 135

Rodeo-Chediski megafire (2002), 31, 141

Safford, Hugh, 149, 243; about, 16, 18–19; on fuel reduction, 22, 240; on mixed-conifer forest fires, 25; on Sierra San Pedro Mártir, 21; on snags, 152

Sauerbrey, Katie, 130–32

Savanna, 111–13

Schulke, Todd, 141–43, 147–48, 229

Scotch broom, 138–39

Seielstad, Carl, 211–12

Serotiny, 70–71

Shortgrass prairie, 112

Sierra Forest Legacy, 147

Sierra Nevada range, 72–73

Sierra San Pedro Mártir, 21

Six, Diana, 85–88, 207

Slash, 143–47

Slow rusting, 213

Smoke, 157, 229–30, 233

Snags, 152–53

Sniezko, Richard, 210–11, 213–15

Snyder, Gary, 3, 93

Soil burn severity, 23, 44–45

Southwestern white pine, 199–200

Spector, Tova, 49, 118–19

Stagnation, 104–6

Stand-removing fire, 23–24

Stand-replacing fire, 23–25

Stephens, Scott, 21, 76–78, 80

Stephenson, Nate, 5, 74–75, 121, 197

Sudden oak death, 178, 212

Sugar pine, 73–74, 197–98

Surface fire, 10, 18

Swetnam, Tom, 99–103, 148–49, 230

Tepley, Alan, 51

Terpenes, 64, 68, 88, 221, 246

Thinning: compared to burning, 127–28; cost, 230; description, 126; in different forest types, 151–52; effects, 22; efficacy, 233; to reduce beetle risk, 81–82

Thomas, Craig, 76, 143, 147, 230, 241

Thomas fire (2017), 213

Timber industry: in 4FRI, 142–43; plantations, 24, 48; reforestation, 48, 175–76; salvage logging, 40, 77; thinning, 128, 150–51

Tomback, Diana: about, 184–85; Clark's nutcrackers, 186, 188, 190; white pine blister rust, 193–94, 202, 206

Tree rings, 10, 93

Trees as a carbon sink, 56–57, 242–45

Tripod megafire (2006), 136, 154

Tropopause, 36

Tyee Creek fire (1994), 50

U.S. Forest Service: 4FRI, 141–42; budget, 77–78, 135, 242; fire suppression, 26, 154–56; prescribed burning, 229; reforestation, 24, 48, 175, 211

Vapor pressure deficit, 15, 121

Verbenone, 68

Vosick, Diane, 143

Wallow megafire (2011), 135, 141

Walsh, Dana, 47–48, 152–53, 176, 213, 240–41

Watershed restoration, 33–34

Watts, Adam, 131–32

Western hemlock, 167

Western larch, 159, 172

Western redcedar, 167–68

Western white pine, 194–95, 196, 199

Whitebark Ecosystem Foundation, 184

Whitebark pines, 183–91, 205–7, 209–12, 241

White pine blister rust: affected trees, 196–202, 224; introduction, 178; introduction and spread, 193–96; range, 202–3; resistance breeding programs, 209–11, 213–15

Wildland-urban interface (WUI), 140, 156–58, 234–35

Yellowstone fires (1988), 55, 71, 96, 158

DANIEL MATHEWS is the author of *Natural History of the Pacific Northwest Mountains*, *Rocky Mountain Natural History*, and *Cascade-Olympic Natural History*. During a career of writing about the natural history of western North America, he has backpacked far and wide, watched for fires from Desolation Peak Lookout, witnessed a forty-inch-thick fir crash onto his family's house in a storm, and lived for several years in a forest cabin without electricity, heating with firewood and writing by kerosene lamp. He lives in Portland, Oregon. Find out more at raveneditions.com.